R-3351-DOE

Terrorism in the United States and the Potential Threat to Nuclear Facilities

Bruce Hoffman

January 1986

Prepared for the U.S. Department of Energy

Rand
1700 MAIN STREET
P.O. BOX 2138
SANTA MONICA, CA 90406-2138

PREFACE

This is the first of two reports prepared under a project sponsored by the Inspector General of the U.S. Department of Energy (DOE) and administered by the Aerospace Corporation. The purpose of the project is to examine the range of potential adversaries to DOE nuclear weapons facilities in the United States and to assess the threat that these adversaries might pose.

A forthcoming report, *A Reassessment of Potential Adversaries to U.S. Nuclear Programs,* R-3363-DOE, by Bruce Hoffman and others, reviews recent trends in international terrorism, violent political protest, criminal activity, sabotage, and malicious mischief that affect the security of U.S. nuclear weapons research sites and production facilities.

SUMMARY

Although the United States is the country most frequently targeted by terrorists abroad, it is near the bottom of the list in terms of the number of terrorist attacks that occur within its own borders. Moreover, despite the fact that the United States has the highest crime and homicide rates in the industrialized western world (and also has the greatest number of both legal and illegal weapons in the possession of its citizenry), politically motivated crimes are relatively infrequent. Nevertheless, three types of terrorist organizations are active in the United States:

- Ethnic separatist and émigré groups.
- Left-wing radical organizations.
- Right-wing racist, anti-authority, survivalist-type groups.

No terrorist group in this country has yet attacked a nuclear weapons facility, and we know of only one threat that has been made to attack such a facility.[1] But this does not imply that U.S.-based terrorist groups have never seriously considered attacking a nuclear weapons facility, or that they may not consider doing so in the future. The available evidence merely indicates that no such plans along these lines have been made. Further, although there may not be a present terrorist threat to U.S. nuclear weapons facilities, the situation could easily change.

The threat posed by domestic terrorist groups is not negligible. The U.S.-based groups most likely to attack a nuclear weapons facility—on the basis of past modus operandi, targeting, motivation, and mindset—are Islamic fanatics, right-wing terrorists, left-wing terrorists, and Puerto Rican separatists. A much smaller threat is posed by Jewish extremists, anti-Castro Cuban militants, Armenian radicals, and anti-abortionists.

The terrorist threat to U.S. nuclear weapons facilities cannot be assessed as high at this time. Terrorists in this country have generally relied on symbolic bombings in pursuit of their political goals. They have consistently avoided attacking *defended* sites, such as nuclear weapons facilities. Thus, if the historical record and perceived inclinations and capabilities of U.S.-based terrorist groups are any guide, it seems unlikely that these groups would attempt such an attack. It is

[1]In April 1984, a then-unknown group, the Radical Nuclear Group, called for the use of "terrorist violence" against nuclear weapons facilities.

not possible to reach the same conclusion with regard to possible state-sponsored terrorism—that is, terrorist groups operating in the United States at the behest and under the direction of a foreign government. State-sponsored groups have greater resources and fewer political constraints, and therefore pose a potentially greater threat to nuclear weapons sites.

ACKNOWLEDGMENTS

I would be remiss not to acknowledge the assistance I received in preparing this report. Karen Gardela deserves special praise for providing superlative research assistance. Geraldine Petty was similarly helpful in steering me to important information that I might otherwise have overlooked, as was Bonnie Cordes in sharing her expertise in Armenian terrorism. I am particularly indebted to Mary Morris and E. Alan Lind, who reviewed an earlier draft of this report. Their detailed comments and suggestions greatly improved the final version. Finally, Janet DeLand brought her invaluable editorial skills to bear in significantly refining an unwieldy manuscript.

CONTENTS

CONTENTS

I. INTRODUCTION

This report examines the threat posed by terrorists to U.S. nuclear weapons production and research facilities. In the last decade, officials of the Department of Energy (DOE) have become increasingly concerned over the likelihood that terrorists might attack a nuclear facility either to gain publicity for themselves and their cause, to create a barricade-and-hostage situation involving facility employees for the purpose of blackmail, to steal strategic nuclear material for their own use or for a patron state, or to sabotage or destroy nuclear weapons or the facility itself.

No terrorist group in this country has yet attacked a nuclear weapons facility, nor do we have hard evidence of any group actually contemplating such an attack. In November 1984, however, police and federal agents raided a Cleveland safehouse used by the radical leftist Sam Melville-Jonathan Jackson Unit[1] and found the name of a nuclear weapons research laboratory on a list of potential targets.[2] We also know of only one threat made against a nuclear *weapons* facility. In April 1984, a previously unknown group, the Radical Nuclear Group, called for the use of "terrorist violence" against nuclear weapons facilities.[3]

Puerto Rican terrorists have twice threatened to attack commercial nuclear *energy* facilities and on one occasion warned that they would detonate several radioactive devices. In March 1984, a leader of the Puerto Rican terrorist organization, the FALN,[4] accompanied by three Mexican members of a radical group, visited the Atomic Weapons Museum in Albuquerque, New Mexico.[5]

It cannot be assumed that U.S.-based terrorist groups have never seriously considered attacking a nuclear weapons facility or may not

[1]This group is discussed in Sec. IV.

[2]See International Association of Chiefs of Police (IACP), Center for the Study of International Criminality, *Terrorist Trends: The Quarterly Intelligence Reporter,* 1985, p. 13. The Brookhaven facility in New York was listed, but federal investigators have since determined that Brookhaven was not a target under "active" consideration. Lists of potential targets were found that contained almost every federal office building and installation in the New York metropolitan area. Some were accompanied by detailed reconnaissance and maps. However, Brookhaven was mentioned with no accompanying description. (Information received from the Office of Safeguards and Security, DOE.)

[3]Information received from the Office of Safeguards and Security, DOE.

[4]*Fuerzas Armadas de Liberacion Nacional* (Armed Forces of the National Liberation). This group is discussed in detail in Sec. III.

[5]IACP, *Terrorist Trends: The Quarterly Intelligence Reporter,* p. 13.

consider doing so in the future. Further, although there may not be a present terrorist threat to nuclear weapons facilities, *the situation must always be considered in a state of flux, with the possibility of sudden changes in the future.*

One problem in assessing the terrorist threat posed to nuclear weapons programs and facilities is that, in the absence of actual terrorist incidents against such targets, evaluations must be based on the historical record of conceptually similar actions. Accordingly, the conclusions of this study are based on an examination of the historical background, motivation, modus operandi, targeting patterns, and mindsets of the known U.S. terrorist groups, and must be considered speculative rather than definitive. Although we can present arguments only in terms of which groups are more or less *likely* than others to undertake such an attack, it is important to keep in mind that a nuclear terrorist incident is nevertheless still quite possible.

II. TERRORISM IN THE UNITED STATES: AN OVERVIEW

Although the United States is the country most frequently targeted by terrorists abroad, it is near the bottom of the list in terms of the number of terrorist attacks within its own borders.[1] Moreover, despite the fact that the United States has the highest crime and homicide rates in the industrialized western world (as well as the greatest number of both legal and illegal weapons in the possession of its citizenry), politically motivated crimes are relatively infrequent[2]. The terrorist statistics compiled annually by the Federal Bureau of Investigation (FBI) indicate that only 29 incidents were recorded in this country in 1980, 42 in 1981, 51 in 1982, 31 in 1983, and 13 in 1984.[3] By comparison, France had 102 reported terrorist incidents in 1980, 59 in 1981, 56 in 1982, and 122 in 1983; Italy experienced 118 terrorist attacks in 1980, 72 in 1981, 26 in 1982, and 9 in 1983; and West Germany recorded 21 acts of terrorism in 1980, 37 in 1981, 28 in 1982, and 6 in 1983.[4] Former FBI Director William Webster stated that the decline in the number of terrorist incidents in the United States was a result of that agency's "increasing success in gathering intelligence about terrorists here and abroad." Webster noted that during the first seven months of 1985, the FBI had thwarted 17 "proposed terrorist incidents."[5]

There are several possible reasons for the relatively low incidence of politically motivated violence in the United States. First, the United States is not a politically polarized country. Unlike France, Italy, or Germany, where a variety of political parties represent the extremes of

---

[1]According to The Rand Corporation's Chronology of International Terrorism, between 1968 and mid-1985, the United States was the country most often targeted by terrorists abroad (1,257 incidents), followed by Israel (293), France (278), the United Kingdom (203), and Turkey (117).

[2]See Bonnie Cordes, et al., *Trends in International Terrorism, 1982 and 1983,* The Rand Corporation, R-3183-SL, August 1984, pp. 2–3.

[3]Federal Bureau of Investigation, Terrorist Research and Analytical Center, *FBI Analysis of Terrorist Incidents in the United States* (see reports for 1981, 1982, 1983, and 1984, published annually in March of the succeeding year). The 1984 figure is somewhat misleading, since attacks on abortion clinics were not included in the FBI count. According to The Rand Chronology of Terrorist Activity in the United States, there were 50 attacks during 1984, of which 21 were abortion-related.

[4]Statistics from Risks International, Inc., 1980–1983.

[5]*The New York Times,* July 9, 1985; and Risks International, *Weekly Risk Assessment,* Vol. 2, No. 29, July 19, 1985.

the ideological spectrum in national politics, the United States has traditionally been a two-party country. Moreover, the two parties differ little in actual substance from one another. Another possible factor inhibiting terrorism in the United States may be this country's unparalleled upward economic and social mobility, which provides opportunities for social and economic advancement. In addition, the United States is a politically absorptive society. Since the turn of the century, American politics have been to a great extent ethnic politics, and immigrants have been readily absorbed by the major political parties and integrated into the American political system. Finally, while other Western nations have violent irredentist groups,[6] there are none in the United States, except for a Puerto Rican faction.

Three types of terrorist organizations, however, do exist in the United States:

- Ethnic separatist and émigré groups.
- Left-wing radical organizations.
- Right-wing racist, anti-authority, survivalist-type groups.[7]

The major organizations within these categories are listed in Table 1.

Approximately 75 percent of all terrorism in this country is carried out by ethnic-separatist or émigré terrorists.[8] Their causes and grievances often have little or nothing to do with domestic U.S.

Table 1

MAJOR TERRORIST GROUPS IN THE UNITED STATES

Ethnic/Émigré	Left-Wing	Right-Wing
Puerto Rican	Weather Underground	Aryan Nations
Jewish	Black Liberation Army	The Order
Cuban	May 19th Communist Organization	Covenant, Sword & Arm of the Lord
Armenian		
Islamic	Melville-Jackson Unit	Posse Comitatus
	Revolutionary Armed Task Force	Anti-abortionists
	United Freedom Front	
	Armed Resistance Unit	

[6]For example, the Irish in the United Kingdom; the Corsicans, Bretons, New Caledonians, and assorted Caribbean separatists in France; and the Basques and Catalonians in Spain.

[7]Anti-authority survivalist groups include some specific, issue-oriented movements of the left as well as the right.

[8]Cordes, et al., op. cit., p. 3.

politics; the United States is simply the battleground where their foreign quarrels are fought. Of the three types of terrorist organizations in this country, the ethnic/émigré groups have generally shown themselves to be the most persistent and violent (e.g., their operations tend to inflict the highest numbers of casualties). These groups also spawn successor generations of younger terrorists. However, despite the potentially wide appeal of these organizations within their own communities, the narrow focus of their parochial, ethnic-centered causes means that they have a far smaller political constituency than ideological terrorist groups have. Support comes only from other ethnic/émigré groups in scattered, tightly knit communities around the country.

Two-thirds of all the terrorist activity in the United States is carried out by five ethnic/émigré movements:[9]

- Puerto Rican separatists
- Jewish extremists
- Anti-Castro Cuban militants
- Armenian radicals
- Islamic fanatics

These five movements by no means encompass all the ethnic/émigré extremists in the country, however. Politically motivated violence has also been perpetrated by Croatian and Taiwanese separatists, as well as by Filipino and Haitian opponents of the Marcos and Duvalier regimes, respectively.

In contrast, left-wing radicals and other issue-oriented groups have a much wider constituency because of the broader political nature of their causes. They are also usually less lethal than their ethnic/émigré counterparts. They engage mostly in symbolic bombings to call attention to themselves and their causes, but they are generally careful not to undertake actions that might alienate potential supporters or their perceived constituency. Although some of the leftist groups justify their existence and operations with vague references to Marxist-Leninist dicta, others are quite specific in their reactions to various contentious political issues.

Radical leftist groups have existed in one form or another since the late 1960s. They originated in student movements that were organized to protest U.S. involvement in the Vietnam War (e.g., the Students for a Democratic Society (SDS)). When the war ended and the U.S. withdrew from Southeast Asia, the related protests and terrorist violence declined. In recent years, U.S. involvement in Central America and

[9]Ibid., p. 4.

South Africa's apartheid policy have given new life to left-wing groups such as the Weather Underground and the Black Liberation Army (BLA). These issues have also led to the creation of new, more narrowly focused, leftist-leaning groups, including the Revolutionary Armed Task Force, the United Freedom Front, and the Armed Resistance Unit.

The right-wing terrorists appear to embrace traits of both the ethnic separatists and left-wing terrorists. They are extremely violent, have no reservations about killing, spawn successor generations, and are often oriented toward specific political issues. Right-wing groups can be divided into specific, issue-oriented terrorists (e.g., anti-abortion crusaders in amorphous entities like the "Army of God") and traditional hate-groups. In recent years, several racist and reactionary groups have surfaced in the West, the South, and the Midwest. Their members include anti-federalists, anti-Semites, racists, survivalists, and Christian fundamentalists. Although related to the Ku Klux Klan and various American Nazi groups, the newer organizations—the Aryan Nations, the Posse Comitatus, the Order, the Covenant, the Sword and the Arm of the Lord—not only champion the old dogmas of a racially pure, Christian United States with no Jews, blacks, communists, Catholics, or atheists, they are also violently opposed to any form of government above the county level, are militantly anti-abortionist, publicly advocate the overthrow of the U.S. government, and in many cases, advocate the extermination of Jews.

The tactic most commonly used by U.S.-based terrorists (like terrorists worldwide) is bombing. Bombings provide a dramatic, but relatively easy means of drawing attention to the terrorists and their causes. Most of the explosions take place at times or in places chosen to cause as few casualties as possible. Emphasis on the use of bombs usually indicates that a terrorist group is at a somewhat "primitive stage" in its development and lacks the organizational expertise, logistics, or knowledge to engage in more complicated or coordinated operations. Few skills are required to manufacture a crude bomb, or to plant one in the dead of night and then be miles away when it explodes.

The emergence of new extremist right-wing groups, however, may herald a new type of American terrorism. The members of these groups are considerably more skilled with weapons than other terrorists in this country. Many, in fact, are accomplished armorers for whom converting semi-automatic weapons to fully automatic machineguns is not difficult. They have large stockpiles of sophisticated weapons, are well-trained in survival techniques and outdoor living, and possess an apocalyptic vision of the future, in contrast to the more neatly defined political aims of the other types of terrorist groups in this country.

III. THE ETHNIC/ÉMIGRÉ TERRORIST GROUPS

Within the five principal ethnic/émigré terrorist group categories, there are a variety of separate, but often interrelated, groups. Table 2 lists these groups and subgroups.

PUERTO RICAN SEPARATISTS[1]

Puerto Rican terrorists have twice threatened to attack commercial nuclear energy installations. In 1979, the FALN (*Fuerzas Armadas de Liberacion Nacional,* or Armed Forces of the National Liberation) threatened to blow up New York's Indian Point nuclear energy facility. Then in March 1980, during the takeover of the Dominican Republic's embassy in Bogota, Colombia, by M-19 (a left-wing Colombian group), the FALN, showing "revolutionary solidarity," warned the United

Table 2

MAJOR ETHNIC/ÉMIGRÉ TERRORIST GROUPS AND SUBGROUPS

Puerto Rican	Jewish	Cuban
FALN	Jewish Defense League	Alpha-66
CRIA	Jewish Armed Resistance	Brigade 2506
FARP	Jewish Action Movement	Cuban Nationalist Movement
EBP (*Macheteros*)	Jewish Direct Action	FLNC
CRP	United Jewish Underground	Omega-7
COPAAN	*Hatikvah Le'umi*	
MAP	Jewish Defense Organization	
OVRP		
Armenian	**Islamic**	
ASALA	Libyans	
JCAG	Iranians	
ARA	Palestinians	
	Black Muslims	
	New World of Islam	
	Fuqra	
	A.M.I.	

[1]For a detailed study of these groups, see William Sater, *Puerto Rican Terrorists: A Possible Threat to U.S. Energy Installations?* The Rand Corporation, N-1764-SL, October 1981.

States: "You must remember . . . that you have never experienced war in your vitals and that you have many nuclear reactors."[2] In addition, a Puerto Rican group (believed to be the FALN) warned in 1975 that it would detonate 100 bombs, of which 25 were alleged to contain radioactive material. Although none of these threats ever materialized, Puerto Rican terrorist groups have in fact assaulted *nonnuclear* energy facilities in Puerto Rico, and one group attempted to bomb a power transmitter in New York.[3]

For more than three decades, Puerto Rican separatists have waged a sporadic, but persistent campaign against U.S. possession of their island. Their goal is to establish an independent and sovereign Puerto Rico. Their first operation, the attempted assassination of President Harry Truman, occurred more than a quarter of a century ago, in 1950. Four years later, separatists attacked the House of Representatives Chamber in Washington, D.C., injuring five Congressmen. Separatist terrorists were then quiescent until 1969, when two groups known as the *Comandos Armados de Liberacion* (Armed Commandos of Liberation) and the *Movimiento de Independencia Revolucionario en Armas* (Armed Independence Revolutionary Movement) began to attack U.S. businesses in Puerto Rico.

The island's police finally dismantled both groups, but their remnants regrouped and united in 1974 to form the *Fuerza Unida Revolucionaria pro Independencia Armada,* or United Force for Armed Independence. This group later became the FALN. The FALN's first operation was the coordinated bombings of five New York City banks on October 26, 1974. For a time, the group continued to carry out symbolic bombings, attacking banks and multinational corporate headquarters—targets the terrorists perceived as representing American capitalism and economic exploitation. The group also bombed banks and a power station in Puerto Rico.

The FALN's most notorious operation occurred on January 25, 1975, when the group detonated an explosive device in the historic Fraunces Tavern near Wall Street during the crowded lunch-hour, killing four persons and wounding 63 others. This target, in the heart of the country's financial center and historically significant in the U.S. war for independence,[4] was undoubtedly chosen for its dual symbolic value.

[2]*Latin,* Buenos Aires, March 21, 1980.

[3]Between 1974 and 1977, Puerto Rican terrorists attacked nine energy facilities in Puerto Rico, including an oil refinery, a power transformer, an electrical substation, electrical transmission towers, and power lines. (Sater, op. cit., App. D, p. 30.)

[4]At the end of the American Revolutionary War, in 1789, George Washington bade farewell to his troops at the Fraunces Tavern.

Although subsequent attacks by the group were apparently planned to avoid causing casualties, the list of targets was expanded to include department stores, hotels, government offices, and military installations, as well as banks and businesses, in New York, Puerto Rico, and Chicago. The most recent FALN operation occurred on New Year's Eve 1983, when a series of bombs exploded at the downtown police headquarters, local FBI offices, a prison, and a federal courthouse in New York City.

The only other Puerto Rican terrorist group that has carried out operations in the continental United States is the CRIA (*Comandos Revolucionarios Independientes Armados,* or Independent Armed Revolutionary Commandos). The CRIA was active only between 1977 and 1978, when it was believed to have been responsible for 20 bombings. Targets included the FBI offices in New York and Newark, New Jersey, and Puerto Rican banks and economic aid offices in New York. Although the bombings caused extensive damage, there were no injuries.

There are a number of other Puerto Rican groups that have not (as yet) carried out any operations in the continental United States.[5] The most active and sophisticated of these is the *Ejercito Borican Popular* (Popular Army of Borica, the island's name before Spanish colonization), or *Macheteros* ("machete-wielders"). Their primary targets are U.S. military personnel and installations in Puerto Rico. In September 1979, the *Macheteros* and the FALN issued a joint communiqué announcing that both groups had joined forces in a common struggle for the liberation of Puerto Rico. The following month, coordinated bombings were carried out in the United States and Puerto Rico. Four bombs planted by the FALN exploded in New York and Chicago; eight others were detonated by the *Macheteros* in Puerto Rico.[6]

The *Macheteros* subsequently embarked on a largely independent campaign of terrorism against U.S. military targets in Puerto Rico. In December 1979, the group ambushed a Navy bus, killing two sailors and wounding ten others; in March 1980, the *Macheteros* attempted to assassinate three Army officers attached to the ROTC program at the University of Puerto Rico; and in January 1981, they destroyed nine jet fighters and caused over $40 million in damages at Muniz Air National Guard Base.

---

[5]These groups include FARP (*Fuerzas Armadas de Resistencia Popular,* or Armed Forces of Popular Resistance), CRP (*Comando Revolucionario del Pueblo,* or People's Revolutionary Commandos), COPAAN (*Comite Patriotico Anti-Anexionista,* or Patriotic Anti-Annexation Committee), MAP (*Movimento de Accion Revolucionario,* or Armed Revolutionary Movement), and OVRP (*Organizacion de Voluntarios de la Revolucion Puertorriquena,* or Organization of Volunteers of the Revolution of Puerto Rico).

[6]*The New York Times,* September 8, 1985.

The most recent *Macheteros* operation occurred in January 1985, when the group claimed responsibility for firing a hand-held antitank grenade at the building in San Juan that housed the offices of the U.S. Marshals Service and other federal agencies. A similar attack against the Federal Building in Hato Rey, Puerto Rico, on October 30, 1983, was claimed by the *Macheteros* as an act of revolutionary solidarity with the people of Grenada. No explanation was offered for the 1985 incident.[7]

The Puerto Rican separatists specifically target symbols of U.S. power and commerce, which they perceive as tools of political and military repression and economic exploitation. All of the separatist attacks in Puerto Rico appear to have been calculated not to inflict casualties among the indigenous population. In contrast, operations in the United States—most notably the Fraunces Tavern bombing—have been far less discriminate and have often deliberately sought to harm innocent bystanders. By targeting military personnel and installations, the terrorists seek to foster the illusion that they are "waging a war of national liberation."[8] In this manner, they (1) focus attention on their cause, (2) publicize their grievances, and (3) "educate" their fellow Puerto Ricans about the alleged inequities forced upon them by U.S. domination and about the need to break the bonds that link their island to the United States.

During the past ten years, Puerto Rican terrorist attacks against people, as opposed to property, have nearly doubled. Further, evidence has come to light that suggests the terrorists have turned to bank robbing and kidnapping to finance their activities. Not only has there been an increase in bank and armored car holdups in both the United States and Puerto Rico,[9] documents seized during a police raid in 1980 revealed that the group had plans to kidnap several American politicians and businessmen—as well as Ronald Reagan, Jr.—for ransom.

The weaponry employed by Puerto Rican terrorists has also become more sophisticated. Whereas earlier attacks employed stolen U.S.

---

[7]Risks International, *Weekly Risk Assessment,* Vol. 2, No. 5, February 1, 1985.

[8]William Sater, "Violence and the Puerto Rican Separatist Movement," *TVI Journal,* Vol. 5, No. 1, Spring 1984, p. 6.

[9]Two of the three operations staged by the *Macheteros* in 1983 involved the robbery of armored cars. (See FBI, Terrorist Research and Analytical Center, op. cit., p. 19.) One of the holdups, that of a Wells Fargo depot in West Hartford, Connecticut, netted the group $9 million. This robbery was committed to coincide with the birthday of Puerto Rican nationalist hero Pedro Albizu Campos. It was planned over the course of 18 months and was the second largest robbery in U.S. history. Thirteen persons suspected of the theft were arrested in Boston, Dallas, Puerto Rico, and Mexico during September 1985. (See *The New York Times,* September 8, 1985; and Risks International, *Weekly Risk Assessment,* Vol. 2, No. 28, September 6, 1985.)

military M-16s or commercially purchased AR-15s, the *Macheteros* used AK-47s in their 1979 attack against U.S. sailors, and they used sophisticated explosives to destroy the nine Muniz jet planes.

The number of Puerto Rican separatist attacks in the United States has been declining. The arrest of 11 FALN members in Illinois in 1980, the capture of FALN bomb expert William Morales in Mexico in June 1983, the arrest of four FALN terrorists in Chicago that same month, and subsequent raids on two FALN safehouses in the Chicago area (that contained "guns, explosives, dynamite, blasting caps, bullet proofs, and disguises") account for this decline.[10] The arrests of 17 key members of the FALN and *Macheteros* in September 1985 may in fact lead to a concomitant decline of terrorist activity in Puerto Rico as well.

THE JEWISH DEFENSE LEAGUE[11]

For more than a decade, the Jewish Defense League (JDL) has been one of the most active terrorist groups in the United States. Although the JDL maintains that it is a political action group concerned with dramatizing the plight of Soviet Jewry and, in more general terms, protecting Jews and Jewish interests worldwide, the FBI has long classified it as a terrorist organization.[12] The vast majority of the group's operations have been directed against Soviet targets, including Soviet diplomatic installations, personnel, and their personal property; Soviet travel and airline offices; Soviet dance troupes and performers; and Soviet and non-Soviet persons and businesses either sponsoring these sorts of cultural events or having trade relations with the Soviet Union.

The underlying purpose of the JDL is to reverse the mythical image of the Jews as victims. This militancy also fuels the anti-Soviet campaign designed to create and foment new sources of tension in Soviet-American relations until persecution of Soviet Jews ends and they are allowed to emigrate to Israel. It deliberately seeks to embarrass the USSR by disrupting Soviet cultural performances and exchanges along with undermining its international trade and commercial activities.

The JDL, however, has also attacked Arab, Iranian, Iraqi, Egyptian, Palestinian, Lebanese, French, and German targets in the United

[10]Risks International, *Weekly Risk Assessment,* September 6, 1985, op. cit., p. 50; and *Los Angeles Times,* August 6, 1985. On August 4, 1985, four FALN terrorists were found guilty of plotting to blow up two military installations in the Chicago area on July 4, 1983.

[11]For a more detailed study of this group see Bruce Hoffman, "The Jewish Defense League," *TVI Journal,* Spring 1984, pp. 10–15.

[12]FBI, *FBI Analysis of Terrorist Incidents in the United States, 1983,* p. 17.

States. In 1976, for example, the Iraqi U.N. Mission was bombed; in 1978, Egyptian diplomats were targeted; in 1981, an Iranian bank in San Francisco was bombed in protest of the treatment of Jews in Iran; and in 1982, a Lebanese restaurant in Brooklyn, alleged by the League to be a PLO headquarters, was set on fire. Attacks have also been staged by League chapters in France, Britain, Italy, and Israel.

The JDL was founded in 1968 in the working- and middle-class Jewish sections of Brooklyn—Borough Park, Crown Heights, and Williamsburg. The religiously orthodox residents of these neighborhoods had long been prey to muggers, vandals, and teenage gangs, primarily groups of young blacks and Puerto Ricans who lived in the ghettos surrounding the Jewish enclaves. Thus the JDL originated as a vigilante street patrol. Within a year of its founding, however, the group became involved in politics as a result of a New York City teachers' strike.

In 1969, the JDL embraced the cause of Soviet Jewry. In early December of that year, a "100-hour vigil" was launched by JDL members who camped outside of the Soviet Union's U.N. Mission in Manhattan. Later that month, the protests escalated; groups of League members simultaneously invaded the offices of the Soviet news agency, TASS, and the Soviet tourist board, Intourist, while a third group handcuffed themselves to an Aeroflot plane which had just arrived at JFK airport. Two years later, the League committed its first act of terrorism. On January 8, 1971, the JDL bombed the Soviet cultural offices in Washington, D.C. An anonymous caller later informed the media, "This is a sample of things to come. Let our people go! Never again!"[13] Since that time, the JDL has pursued a dual-track strategy of acts of civil disobedience and generally peaceful protest, along with acts of outright terrorism.

In 1971, the JDL's founder, Rabbi Meir Kahane, and his family left the United States and settled in Israel. Within a few months, an "International JDL Office" was opened in Jerusalem. Although Kahane commuted back and forth between Israel and the United States, his absence greatly affected the American branch of the JDL and it gradually fell into disarray. In August 1985, Kahane suddenly announced that he was stepping down as leader of the JDL and had appointed Irv Rubin in his place. Rubin, the JDL's West Coast coordinator, is a private investigator and sometime painter, who boasts of having been arrested 30 times in connection with Jewish causes. The ostensible reason for Kahane's resignation was his desire to devote more time to his parliamentary duties in Israel (he was elected to office

[13] *The New York Times,* January 9, 1971.

in 1984) and to his fledgling *Kach* party.[14] But the real reason may have been Kahane's desire to distance himself *publicly* from the intensified terrorist campaign now being waged by the JDL and thus avoid compromising his efforts to obtain contributions from American Jews for his campaign to become Israeli Prime Minister.

In February 1985, the JDL took credit for a bomb discovered beneath a car belonging to a Soviet diplomat in Manhattan (the bomb was defused by bomb-squad technicians).[15] The JDL is also believed to have been responsible for a bomb that was found in front of a Los Angeles cinema presenting a Soviet film festival in April. That device too was dismantled by the police. The following month, the JDL claimed credit for bombing the house of George Ashley, a Los Angeles-area high school teacher, who had told students that the number of Jews killed by the Nazis during World War II was considerably less than the commonly accepted figure of six million, and that not more than a million Jews perished during the war.[16] Then, in June, an anonymous caller stated that the JDL had intentionally set fire to the offices of the German-American National Political Action Committee in Santa Monica, California. The German-American group probably incurred the JDL's wrath by its advocacy of the view that "Jews were not gassed by the Nazis . . . [and that the] numbers and reports of predetermined extermination are greatly exaggerated by professional liars."[17] In each of these cases, official JDL spokesmen subsequently denied or disavowed any connection between the group and these attacks.

In August 1985, the locus of alleged Jewish terrorist activities shifted to the northeastern United States. On August 16, a caller claiming to represent the JDL stated that the group had planted bombs that were discovered in various places around West Roxbury and Cambridge, Massachusetts (the West Roxbury device exploded as police attempted

---

[14] *Los Angeles Times,* August 20, 1985; and *Washington Post,* August 31, 1985.

[15] Risks International, *Weekly Risk Assessment,* Vol. 2, No. 8, February 22, 1985. Exactly a year earlier, the residential compound housing the families and members of the Soviet Union's U.N. Mission in the Bronx was bombed. That attack was claimed in the name of Jewish Direct Action (JDA), although a group spokesman later denied responsibility. The composition, and even the existence, of the JDA is unknown. It may be a *bona fide* splinter group from the JDL or it may exist only on paper. In any case, the JDL traditionally carries out at least one operation in February, around the time of the Jewish festival of *Purim.* This holiday commemorates the ancient liberation of the Jews from enslavement in Babylonia—an obvious parallel to the treatment of present-day Soviet Jewry.

[16] *Los Angeles Times,* May 8, 1985.

[17] *Los Angeles Times,* June 6, 1985.

to dismantle it). Another bomb was found in front of the offices of the American-Arab Anti-Discrimination Committee (ADC) in Boston. Two police officers were injured trying to disarm it. The JDL, however, denied that its members had any part in the incident.[18] That same day, an explosive device rigged to the front door of the Patterson, New Jersey, home of Tscherim Soobzokov, who was once accused by the U.S. Justice Department of concealing his wartime membership in the Nazi Waffen-SS, exploded. Soobzokov was critically wounded and died three weeks later. Less than a month later, a similar type of bomb exploded outside the home of another alleged Nazi war criminal in the Brentwood section of Long Island, New York. Although the target of the attack, Elmars Sprogis, a former Latvian policeman, was not hurt, a passerby was seriously injured. Shortly after the bombing, a local newspaper received a call from a person purporting to speak for the JDL, claiming credit for the explosion.[19]

The most recent incident of suspected Jewish terrorism occurred on October 11, 1985, when the office of the ADC in Santa Ana, California, was bombed. The Committee's director, Palestinian-born Alex Odeh, was killed by the bomb, and eight passersby were injured. The attack came at the end of a long week which had seen the seizure of an Italian luxury liner, the *Achille Lauro,* on the Mediterranean Sea by Palestinian terrorists; the murder of an American passenger, Leon Klinghoffer; the negotiated release of the crew and passengers; and the in-flight interception by U.S. Navy fighters of an Egyptian airliner carrying the four hijackers of the ship. Although Odeh was accustomed to receiving anonymous death threats, this attack seemed to have been prompted by his appearance on a local news show the previous evening, in which he described Yasir Arafat as "a man of peace." [20] It was also speculated that Odeh was targeted in revenge for the murder of Klinghoffer, a Jew.

Bombing is the JDL's favored tactic, accounting for 78 percent of all JDL incidents. Shootings are next, accounting for 16 percent, followed by arson, vandalism, and kidnapping, accounting for 1 percent each. In the past, the preponderance of bombings suggested that the JDL was still at a relatively primitive stage and was not able to undertake more sophisticated operations. However, the group's technical proficiency has apparently increased appreciably. The new generation of recruits joining the group may have greater knowledge and sophis-

[18]Risks International, *Weekly Risk Assessment,* Vol. 2, No. 34, August 23, 1985.

[19]Ibid., September 7, 1985.

[20]*Los Angeles Herald Examiner,* October 13, 1985; and *Washington Post,* October 13, 1985.

tication in terrorist tactics in general and the manufacture of bombs in particular. This sophistication may also be product of the summer camps run by the JDL in New York and California for Jewish youths. A leaflet issued by the group describes one camp, known as JUDO (Jews United to Defend Ourselves), as a place that offers the opportunity to "learn Karate, legal and proper weapons training, street fighting and how to be a proud Jew who can defend the Jewish people."

Since 1968, JDL operations have killed 7 persons and wounded at least 22. Thirty-nine percent of the targets were connected with the Soviet Union; 9 percent were Palestinian; 8 percent were Lebanese; 6 percent, Egyptian; 4 percent, French, Iranian, and Iraqi; 1 percent, Polish and German; and 23 percent were not connected with any state. Sixty-two percent of all JDL attacks are directed against property; 30 percent against businesses; 4 percent against academics and academic institutions; and 2 percent against religious targets.

Typically, an anonymous caller will claim responsibility for a specific terrorist act for either the JDL or one of its alleged subgroups, only to have an official spokesman for the JDL deny the group's involvement the following day.

In the past, although the JDL was among the most active terrorist organizations in the United States, the threat it posed appeared to be primarily symbolic. The JDL used terrorism to draw attention to itself and its causes, to maintain momentum, and essentially to stay alive. Recent events, however, suggest that this view requires revision. The increase of militant Jewish terrorism represents not only an escalation of violence but a significant change in targeting patterns, as well as a dramatic shift in tactics.

Nine times as many terrorist acts have either been claimed by or allegedly carried out by the JDL in 1985 as in either 1983 or 1984 (one incident occurred in each of the two earlier years). Whereas Soviet targets were previously the terrorists' primary targets, only two attacks in 1985 were connected with the Soviet Union. Hence, it now appears that nonstate targets in the United States that are perceived by the terrorists as hostile to Jewish interests have become their primary focus.

As far back as 1976, the JDL had expanded its scope of activities to include Iraqi, Iranian, Egyptian, Palestinian, Lebanese, French and German targets. But these operations were isolated and infrequent and were consistently overshadowed by the group's preoccupation with the issue of Soviet Jewry. Now, however, the group appears to be concentrating its efforts on persons and institutions it considers to be enemies of Judaism and Israel. The targets now include alleged former Nazis and war criminals; Palestinian and Arab individuals and institutions;

and persons and so-called research centers promoting views about the Holocaust that minimize the dimensions of Jewish suffering.

Perhaps the most far-reaching change, however, is the increasing use of assassination, both to draw attention to the terrorists' causes and to eliminate perceived enemies of the Jewish people and Israel.

This is an ominous development. Political violence does not occur in a vacuum, and it is responsive to both internal and external stimuli. In the case of the Jewish terrorists, the internal stimulus undoubtedly is the decline in membership and financial contributions that their groups have experienced in recent years. By expanding their targets, they may also be trying to appeal to a larger and more diverse constituency. This could represent a bid to regain momentum and duplicate the JDL's successes of the 1970s.

External stimuli probably include the growing sense of unease felt by Jews in the United States as a result of the anti-Semitic insinuations and proclamations issued by the Reverend Louis Farrakhan, leader of the Black Muslim Nation of Islam,[21] the growing evidence of a widespread network of right-wing and neo-Nazi extremists,[22] the hijacking of a TWA aircraft in June 1985 by radical Shi'as in which Jewish passengers were separated from gentiles, and the murder of Leon Klinghoffer. All of these developments have alarmed Jews, and more important, they may have convinced the terrorists that the Jewish public has become more amenable to and supportive of militant Jewish activity.

The increase in Jewish terrorist activity may prove to be a harbinger of continued escalation of such activity in the United States, the use of increasingly violent tactics, and even, perhaps, an expansion of the terrorists' already broadened range of targets.

THE CUBAN ANTI-CASTRO TERRORIST MOVEMENT[23]

Since 1968, Cuban exile terrorist groups based in Miami, Florida, and Union City, New Jersey, have fought an unrelenting and bloody war against the Castro regime. Their long-term goal is to overthrow the Castro regime and end Communist domination of their island homeland; their immediate objective, however, is to disrupt diplomatic, commercial, and cultural relations between Cuba and other countries. Their base of popular support within the large Cuban émigré communi-

---

[21]Farrakhan and the Nation of Islam are discussed in detail below.

[22]Right-wing terrorists are discussed in detail in Sec. V.

[23]For a detailed examination of this movement, see Bruce Hoffman, "The Cuban Anti-Castro Terrorist Movement," *TVI Journal*, Spring 1984, pp. 15–22.

ties in the United States has provided the organizational capacity, manpower, expertise, and arsenal required to sustain their struggle. As a consequence, anti-Castro terrorists have been able to conduct a long-standing campaign of violence against the Cuban government and those states, commercial interests, and persons that have relations or trade with Cuba.

The major Cuban terrorist groups are Alpha-66, Brigade 2506, the Cuban Nationalist Movement, the FLNC (*Frente de Liberacion Nacional Cubana,* or Cuban National Liberation Front), and Omega-7. Of these, only Omega-7 has been active in the United States during the past five years. In July 1983, Eduardo Arocena, the leader of the group, was arrested. During his arraignment, it was revealed that Arocena had been an FBI informant for over a year. He had decided to inform on his compatriots because he believed that they had become "Communist sympathizers" and were more interested in trafficking in drugs than in prosecuting the war against Castro. By turning in the other members of Omega-7, Arocena hoped to "cleanse" his organization. In fact, he virtually destroyed it.

The general strategy of the anti-Castro terrorist movement has been the disruption of Cuban diplomatic and commercial activities, the thwarting of progress toward the "normalization" of relations between Cuba and other countries, and the intimidation of those states, businesses, and individuals who trade with Cuba, arrange travel or cultural exchanges with it, or are sympathetic toward the Castro government.

All but 19 percent of the Cuban terrorists' operations have been staged in the United States. Cuban interests and Cuban diplomats are the most frequent targets, followed by members of the Cuban exile community in the United States. The types and nationalities of targets have varied only slightly: 38 percent of the targets have been embassies, consulates, and diplomatic missions; 27 percent, business concerns; 16 percent, offices of various countries' national airlines; 9 percent, ships and tourist offices; 2 percent, Communist and Socialist party offices; and 1 percent, banks. Bombings account for 90 percent of the attacks.

The efforts of the anti-Castro Cubans have been generally unproductive: Cuba has not been liberated; Castro remains firmly entrenched in power; and the terrorists have been ineffective in disrupting diplomatic, trade, and cultural relations between Western countries and Cuba. As the goal of overthrowing Castro has become increasingly distant and the efforts of the terrorists to liberate Cuba have brought no tangible results, the political goals of the anti-Castro movement have been increasingly subverted by personal avarice. Nevertheless, it cannot be presumed that the threat in the United States posed by the Cuban terrorists will necessarily disappear. Past events have illustrated their organizational resiliency and their ability to rebound from setbacks, adapt to rapidly changing circumstances, and carry on their struggle.

THE ARMENIANS[24]

Armenian terrorists have carried out more than 200 attacks since their emergence in 1975, but only six of these have been staged in the United States. Indeed, there have been none in this country since 1982. The Armenian terrorists' targets in the United States have been mostly limited to Turkish diplomats (and their families), embassies and consulates, airline and tourist offices, and folk-dance troupes. The Armenian groups have rarely carried out indiscriminate bombing or shooting attacks in public places.

The three main Armenian terrorist groups are the Armenian Secret Army for the Liberation of Armenia (ASALA), the Justice Commandos for the Armenian Genocide (JCAG), and the Armenian Revolutionary Army (ARA), an offshoot of the JCAG. Although their tactics and choice of targets differ somewhat, all three share the same goals: revenge for the events of 1915 (when the Armenian inhabitants of eastern Turkey were massacred); Turkish recognition of the genocide; and reparation payments from Turkey to the survivors and their families. To this end, 42 Turkish diplomats and members of their families have been assassinated since 1975. The terrorist groups are highly secretive, difficult to penetrate, and held together by strong ethnic bonds. The real and perceived sympathy emanating from the largely law-abiding Armenian community both in the United States and abroad is a crucial part of the terrorists' determination to carry on with their struggle.

The JCAG and the ARA restrict their attacks to Turkish targets, carrying out either discriminate, symbolic bombings (which rarely cause injury), assassinations of specific Turkish officials, or seizures of Turkish embassies and consulates. The ASALA uses these tactics and attacks the same types of targets, but has been far more indiscriminate, causing large numbers of civilian casualties among Turks and non-Turks alike.[25] The ASALA has also carried out bombing campaigns against countries that arrest and incarcerate members of the group.

---

[24]For a detailed examination of this movement, see Bonnie Cordes, "Armenian Terrorism in America," *TVI Journal*, Spring 1984, pp. 22–27.

[25]For example, in July 1983, the ASALA bombed Paris's Orly Airport, killing seven persons and wounding 56 others; a machinegun and hand-grenade attack at Esenboga Airport in Ankara a month later killed nine persons and injured 78; and a similar assault in Istanbul's Grand Bazaar killed two persons and wounded 27.

The first Armenian terrorist attacks in the United States occurred in October 1980, when the ASALA claimed credit for bombing attacks at the Turkish Mission to the U.N. in New York and a Turkish-owned travel agency in Hollywood, California, while a JCAG member fire-bombed the Bel Air, California, house of the Turkish consul-general. Three further bombings were mounted by ASALA terrorists against Swiss interests. Switzerland was targeted by the group in reprisal for the arrest and imprisonment of several ASALA members in that country. In June 1981, a bomb was planted by the JCAG at the Anaheim (California) Convention Center to disrupt a performance of a Turkish folk-dance troupe. The explosion did, in fact, lead to the cancellation of the performance. Six months later, another bomb caused extensive damage to the Turkish consulate in Beverly Hills on the eve of the visit of Turkey's ambassador to the United States. In May 1982, the ASALA bombed the Swiss Bank Corporation in New York, and three terrorists were arrested for attempting to blow up an Air Canada cargo building at Los Angeles International Airport in retaliation for the arrest a few days earlier of alleged Armenian terrorists in Toronto on charges of attempting to extort money from law-abiding Armenian-Canadians to finance the ASALA's struggle.

The first terrorist assassination of a Turkish diplomat in the United States took place in January 1982, when Kemal Arikan, the Turkish consul-general in Los Angeles, was shot to death by two members of the JCAG. Then in May 1982, Orhan Gunduz, the honorary Turkish consul-general in Boston, was gunned down by a lone assailant. Once again, the JCAG claimed responsibility. Gunduz's assassination came only weeks after his shop had been bombed. Although this was the last Armenian terrorist operation actually perpetrated in the United States, in October 1982 police arrested five young Armenians in Boston en route to Philadelphia to blow up the office of the Turkish honorary consul, Kanat Arbay. The absence of any further Armenian terrorist operations in this country is probably the result of the harsh prison sentence handed down to one of Arikan's assassins by a California court.[26]

ISLAMIC FANATICS

On December 13, 1983, just two days after the bombing of the American embassy in Kuwait, Iran's ambassador to the United Nations, Said Rajaie Khorassani, spoke of the continuing U.S. presence in

[26]Hampig Sassunian received a life sentence with no possibility of parole.

Lebanon: "The U.S. forces have bombed not only military bases but civilians. The United States is therefore openly a party to conflict. It must prepare itself for further punishment from the Lebanese Muslims and consequently [for] further retaliation against innocent people."[27] The possibility that Islamic terrorists might commence operations in the United States has led to the implementation of unprecedented physical security measures at government offices and military installations throughout the country. Concrete pylons and barriers were installed at the White House, the Department of State, the Pentagon, and other buildings in Washington, D.C., and dump-trucks blocking driveways and entrances to underground garages have become an everyday sight. Although suicide car-bomb attacks have not materialized in this country, the threat of Middle Eastern terrorist violence has not abated.

It is feared that Islamic fundamentalist terrorists have already built a terrorist infrastructure in this country. For example, according to the Institute of International Education, there are some 25,000 Iranians studying in U.S. universities.[28] Although there is no evidence to suggest that these students are in fact terrorists or minions of the Khomeini regime, at least some of them could be "agents-in-place," trained in terrorist tactics and awaiting orders to begin operations.

Following a conference of Iranian clergymen in Teheran in March 1982, for example, the Khomeini regime reportedly established a special program to train Muslims from every nation of the world as "messengers of true Islam." After receiving instruction in Iran, these militants were to return to their home countries to foment unrest and create a climate that would favor the adoption of fundamentalist Islamic precepts. Whether or not similar "messengers" were dispatched to non-Islamic countries in the West to undertake subversive, sabotage, and terrorist activities at the behest of Khomeini is unknown. It is clear, however, that Iranians residing in the United States have been involved in the assassination and intimidation of opponents of the Khomeini regime.

There are two known pro-Khomeini groups presently active in the United States: the Islamic Guerrillas in America, headed by a naturalized Iranian citizen living in Washington, and the Moslem Students Organization. Both groups have been linked to efforts directed by Teheran to stifle Iranian critics of Khomeini. On at least two occasions, intimidation has given way to assassination. In July 1980, Ali Tabatabai, who served as press attaché for the Shah of Iran before the 1979 revolution that brought Khomeini to power, was murdered in

[27] *The New York Times,* December 14, 1983.

[28] Ellen Weiss, "Islamic Terrorist Threat in U.S.," *TVI Journal,* Summer 1984, p. 28.

Washington, D.C., by David Belfield—a Black Muslim also known as Daoud Salhudin—who is a member of the Islamic Guerrillas. The assassination was allegedly carried out on behalf of the Iranian government. Belfield later fled the United States and is believed to have gone to Iran.[29] Less than two weeks later, another Iranian dissident, Shah Reis, was shot to death in Los Angeles.[30] American Black Muslims were allegedly recruited to carry out the murder.[31] The Islamic Guerrillas, in fact, recruit many of their members from the District of Columbia's black prison population.[32]

U.S.-based Libyan groups, such as the People's Committee for Students of Libyan Arab Jamahriya and the Libyan Student Center, have also been implicated in attacks on opposition figures living in this country. Libyan leader Mu'ammar Qaddafi has frequently proclaimed that dissidents living outside of Libya will be pursued and punished for their treasonous acts. On at least two occasions, this threat was carried out in the United States. In October 1980, a Libyan graduate student at Colorado State University who had criticized the Qaddafi regime was seriously wounded in an attempted "contract" assassination. The would-be assassin was a former U.S. Special Forces "Green Beret" hired by another American employed by the Libyan government. The following July, another Libyan student was found murdered in Utah. A fellow student—also a Libyan national—was arrested as he attempted to flee the United States and return to Libya. He was subsequently sentenced to prison. In August 1984, two Libyan students were arrested while attempting to purchase silencer-equipped firearms from undercover FBI agents.

More recently, a federal grand jury was impaneled in June 1985 to investigate charges that expatriate Libyans in this country had plotted to assassinate at least three Libyan opponents of the Qaddafi regime. Some 15 persons were subpoenaed to appear before the jury, four of them employees of the Libyan Student Center near Washington, D.C. The Center was established to help Libyan students in this country after Libyan diplomats were expelled from the United States in 1981.[33] Whether by coincidence or not, shortly after the grand jury began to hear testimony, a Libyan diplomat attached to his country's U.N. dele-

---

[29] *Washington Times,* June 26, 1984.

[30] IACP Center for the Study of International Criminality, *CTT: Terrorist Trends—The Quarterly Reporter,* 1985, p. 8.

[31] Weiss, "Islamic Terrorist Threat," p. 28

[32] *Washington Times,* June 26, 1984.

[33] Risks International, *Weekly Risk Assessment,* Vol. 2, No. 23, June 7, 1985.

gation was deported on grounds of his involvement in planning the assassination of Libyan dissidents here.[34]

Close bonds have been established between the Nation of Islam, an American Black Muslim Sect with about 10,000 members, and Qaddafi. In February 1985, Qaddafi addressed the religious movement's annual convention in Chicago via a satellite hook-up. Qaddafi called on Black servicemen in the U.S. Army to leave the military, create their own army, and fight "your racist oppressors" to establish a separate Black nation in this country. The United States, he declared, is an "imperialist country [that] must be destroyed. . . . You have the force, you have the soldiers. We are ready to give you arms because your cause is just. . . . We will fight together shoulder to shoulder." The Reverend Louis Farrakhan, leader of the Nation of Islam, was described by Qaddafi as "a fellow struggler in the cause of liberation." In response to Qaddafi's entreaties, Farrakhan stated, "It would be an act of mercy to end the white man's world because your world is killing you and us and all of humanity," although he later qualified his statement, pointing out that he was "not trying to overthrow the U.S."[35] Three months later, Farrakhan announced that he had received a $5 million, interest-free loan from Libya. He declared, "I am not a terrorist and don't you say I've taken this money to make bombs." Farrakhan said that the money would be used to promote economic independence among blacks.[36]

The New World of Islam, another black Muslim group, has been directly implicated in acts of terrorism in this country. The group, which advocates the establishment of a separatist black nation in the American south, has staged at least 21 bank robberies in the New York and New Jersey metropolitan areas since 1978. Through such robberies, the New World of Islam hopes to finance its independent homeland. Fanciful though this goal may seem, in May 1985 the New York Police Department issued orders to its personnel advising them of the seriousness of the group's threat to "execute" any police officers they encounter.[37]

There is also a fear that Palestinian terrorists may be poised to begin operations in the United States. Unconfirmed Israeli intelligence reports, for example, point to a PLO Central Committee meeting held in late 1982 where this idea was raised. Although the moderates were able to prevent the implementation of such a plan, the fact that the

[34]Ibid., No. 24, June 14, 1985.

[35]*Washington Post,* February 25, 1985.

[36]*The New York Times,* 4 May 1985.

[37]*The New York Times,* May 11, 1985.

matter had been raised, even if it was not seriously contemplated, was considered an ominous development.[38]

Apparently, a number of Palestinian residents in this country have traveled to the Middle East to receive military training. These people have now returned to the United States, and although there is no evidence that this training was meant for future terrorist activities, the training *may* have been part of a plan to establish some sort of cadre for terrorist assignments here. In July 1983, the FBI broke up a plot by members of an undisclosed faction of the PLO to attack a fellow Palestinian belonging to a rival faction.[39]

Finally, there is an internecine rivalry among domestic Muslim factions, primarily between the *Fuqra* (Truth) sect and the Ahmaddiya Movement in Islam (AMI). *Fuqra* seeks to further its fundamentalist religious goals by eliminating rival Islamic sects. In the past, its adherents have employed assassination, firebombing, and arson to to achieve this end. In 1983, *Fuqra* staged three attacks (including the firebombing of the house of the AMI treasurer), killing three AMI members.

THE ETHNIC/ÉMIGRÉ TERRORIST THREAT TO U.S. NUCLEAR WEAPONS PROGRAMS

Although ethnic/émigré extremists have been the most active terrorists in the United States, there is strong evidence that this may be changing. The diminution of this activity is less the result of a decrease in political violence than the gradual demise and dissolution of many of the organizations. Two of the principal groups—the Puerto Rican separatists and the anti-Castro Cubans—have sustained serious losses to both their leadership cadre and rank-and-file membership as a result of government successes and arrests. Armenian terrorist operations in this country have also stagnated because of the arrests of several Armenian terrorists and the harsh prison sentences handed down to them. Islamic political violence has remained at a sporadic level. Only Jewish-related terrorism has increased.

The decreasing activity of some groups may also be indicative of a substantial drop in membership, or at least of a decline in the number of recruits joining the organizations. This in itself is both out of character for these groups and surprising. Ethnic/émigré terrorist organizations, though lacking as wide a constituency as ideological or issue-

[38]Information obtained through personal communication with Israeli intelligence sources.

[39]FBI, *FBI Analysis of Terrorist Incidents in the United States, 1983*, p. 47.

oriented groups, have always been able to replenish their ranks with new recruits from younger generations. The inability of the Puerto Ricans, Cubans, and Armenians to carry on their respective struggles following large-scale arrests of their members may indicate that recruiting has become a problem. In the Puerto Rican and Cuban cases, it may be that as time has dragged on and the terrorists' goals have not been realized, fewer young people are stirred to join in the struggle. By the same token, the factors that have led to an increase in Jewish terrorist activity may have also created a new generation of terrorists. The Islamic terrorists are an unknown—but potentially ominous—element in this context.

The net cast by Islamic terrorists is broad, because of the diverse causes and nature of the groups in this category. Groups composed of pro-Qaddafi Libyan and pro-Khomeini Iranian nationals resident in the United States, Palestinians, and Black American Muslims have targets ranging from expatriate dissidents to Israeli and Jewish interests to competing religious sects or terrorist groups. Nevertheless, the Islamic terrorists' pattern of operations has remained remarkably constant—centering primarily on the intimidation and assassination of fellow countrymen or members of rival religious or political organizations.

If present trends of Puerto Rican separatist terrorism continue despite the recent wave of arrests, U.S. government property and military personnel will remain the principal targets of these groups. Targets will continue to include police stations, federal courthouses, government buildings, and military personnel and installations. Banks and other commercial establishments as well as energy facilities are also likely targets of future operations.

There is no reason to assume that the Puerto Rican terrorists will make good on any of their past threats to attack either a nuclear energy facility or a nuclear weapons site, or to employ a nuclear weapon. The tactics and weapons used by Puerto Rican terrorists (except for the *Macheteros*) have not been particularly sophisticated, nor have operations against well-defended targets been common. Thus is appears that the Puerto Rican terrorists would have a great deal of difficulty staging a successful attack against a U.S. nuclear weapons facility.

The JDL's ideology and choice of targets would seem to preclude any serious consideration of an attack on a U.S. nuclear weapons facility. Not only would it be thoroughly out of character, it would have no appeal to the JDL's perceived constituency and no practical significance in advancing the JDL's political goals.

The Cuban exile terrorists' heavy involvement in the drug trade has curtailed their *politically oriented* activities. Hence, there is little likelihood that they pose a threat to U.S. nuclear weapons facilities. It should be noted, however, that the anti-Castro Cubans are the best trained of the terrorist groups in the United States and the most proficient in the use of weapons and in paramilitary tactics. Their past willingness to hire themselves out as "hit-men" and "enforcers" for drug dealers, coupled with their weapons and operational expertise, may suggest that these well-trained terrorists could serve as mercenaries for a government or a private individual intent on stealing nuclear material. It is also entirely conceivable that a nuclear weapons site could be seized for blackmail purposes. But an exile terrorist attack on a facility for some political/publicity purpose would be out of character.

The potential threat posed by the Armenian terrorists to U.S. nuclear weapons facilities is negligible at best. Such an operation would not only be out of character, it would be totally unrelated to the Armenian cause and would contribute nothing toward achieving the Armenians' goals.[40] It would certainly alienate the Armenian community in the United States and would diminish, rather than increase, support for the terrorists.

The terrorist activities of Islamic extremists in this country have largely been confined to assassinations of specific individuals and the intimidation of members of rival Muslim sects. While the latter has no relation to the security of U.S. nuclear weapons facilities, the former does, since it presents solid evidence of state-supported terrorism. Although Libyan and Iranian involvement in terrorism in the United States does not as yet constitute a serious threat, the presence of persons who are willing to undertake terrorist missions on behalf of the Qaddafi and Khomeini regimes is a matter of considerable concern—particularly in view of the Libyan "hit-squads" allegedly sent to the United States in December 1981 to assassinate President Reagan. Islamic extremists in the United States who have tight links (religious, financial, and ideological) with Libya and Iran could be employed as surrogates, as mercenaries, or in cooperation with foreign agents infiltrated into this country to attack a U.S. nuclear weapons facility.

---

[40]A previously unknown organization, the Armenian Scientific Group, did send a letter to a Belgian news agency warning that it had three small nuclear bombs which it would use to destroy Istanbul, Ankara, and Izmir if the "Armenian genocide resolution" before the United Nations was not endorsed by Turkey. The letter was written in "very bad English" and was "not taken very seriously." (See *Mamara,* Istanbul, January 14, 1985.)

IV. LEFT-WING TERRORIST GROUPS

In recent years, there has been a resurgence of left-wing terrorism in the northeastern United States. Issue-oriented groups, dormant since the end of the Vietnam War, have reemerged to champion new causes, and new groups have formed. These new groups have claimed responsibility for 16 bombings in the New York and Washington, D.C., metropolitan areas during the past four years. In New York, multinational corporations, military facilities, and government offices have been the primary targets. Attacks have been carried out against branch offices of IBM, Honeywell, Motorola, Union Carbide, and General Electric, as well as the local FBI office, a Navy recruiting station, a Naval Reserve Center, and a National Guard armory. In Washington, D.C., the Senate wing of the U.S. Capitol was bombed, as were the Navy yard and a Navy recruiting station.

Although the exact composition of the new groups is not known, law enforcement personnel believe that all of these organizations are in some way related to the Black Liberation Army (BLA). They base this conclusion on several facts:

1. The wording of communiqués issued by the new groups is extraordinarily similar to that used by the BLA.
2. Identical types of weapons are employed by the groups.
3. The new groups and the BLA generally select targets of similar symbolic significance.[1]

The major indigenous left-wing terrorist groups are shown in Fig. 1.

The leftist terrorists can generally be described as "uniformly anti-U.S., anti-imperialist organizations that believe that the root causes of the world's ills lie in America's exploitation of non-White Third World peoples."[2] Their targets—banks, corporate offices, and military facilities—are chosen to publicize the terrorists' cause and existence, as well as to symbolize their anti-imperialist/anti-capitalist ideology. Operations are staged to generate what the terrorists refer to as "armed propaganda."[3] The rationale behind this use of terrorism is explained by Thomas Manning, a member of the United Freedom Front:

---

[1] *Wall Street Journal,* July 26, 1984.

[2] IACP, *Terrorist Trends,* p. 11.

[3] IACP, *Clandestine Tactics and Technology,* Vol. IX, Issue 2, p. 2.

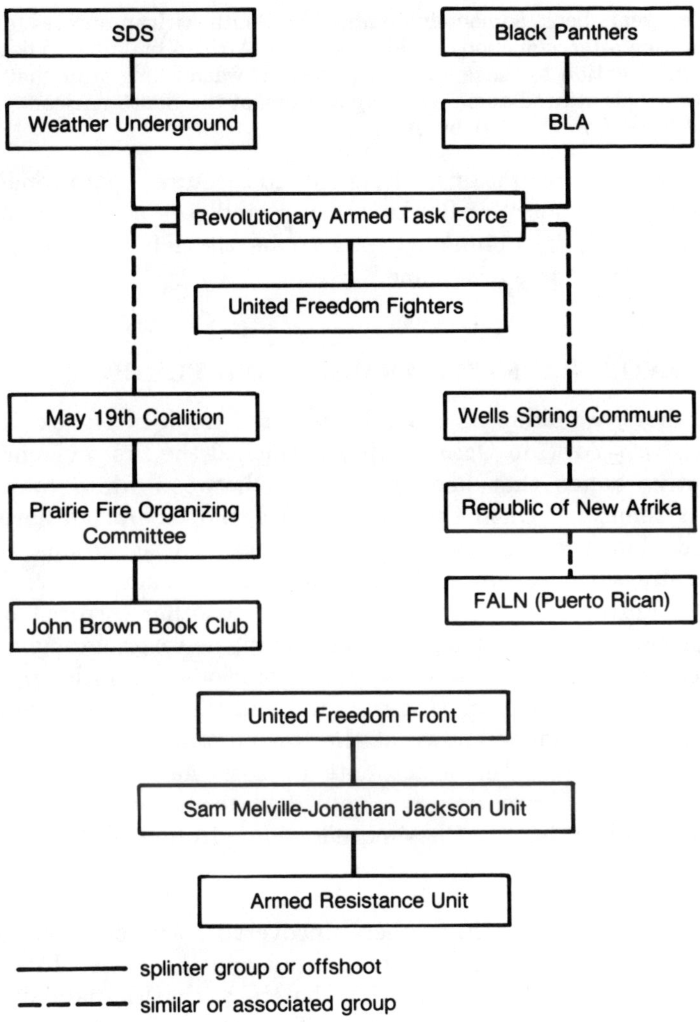

Fig. 1—Major left-wing terrorist groups

> A few years back, somebody bombed the South African airlines office. Not long after, somebody bombed a South African building. That draws a lot of attention to those kinds of places. It wasn't long after that that people started demonstrating in front of the South African embassy—now that's popular.[4]

Other types of operations, designed to acquire operational funds, include armed robbery of banks and armored cars as well as narcotics trafficking.[5] Bank robbing appears to be the leftists' principal, and most lucrative, source of income.[6]

THE REVOLUTIONARY ARMED TASK FORCE

The abortive holdup of a Brinks armored truck in Nyack, Rockland County, New York, in October 1981 furnished the first evidence of the cooperative bonds that had been forged between white radicals and black extremists.[7] Until that time, police and federal authorities had assumed that the variety of small, radical leftist groups that had periodically surfaced since the end of the Vietnam era were wholly separate and distinct, with no connection to other terrorist organizations in this country. The Brinks robbery, however, brought to light evidence that two of America's most notorious revolutionary groups, the largely white Weather Underground and the exclusively black BLA had formed an alliance known as the Revolutionary Armed Task Force (RATF).[8] Although the exact date of this merger is unclear, it is believed that the two groups established formal relations sometime after BLA leader Joanne Chesimard's escape from a New Jersey prison in 1979.[9]

The RATF was responsible for the Brinks robbery, and police raids on safehouses used by the robbers uncovered documents that revealed the merger of the Weather Underground and the BLA. In itself, the merger was unprecedented. During the era of the Vietnam War, white leftist radicals and black militants with similar ideological orientations were unable to establish cooperation. Despite the radicals' efforts, their black counterparts tended to resent the whites' affluent suburban backgrounds and disparaged them for exhibiting "revolutionary

[4] *USA Today,* July 15, 1985.

[5] Ibid., pp. 4–5; and Martin C. Arostegui, "Terrorism in the United States," *Clandestine Tactics and Technology: Group and Area Studies,* Vol. IX, p. 4.

[6] Statement made by FBI agent Manuel Marquez, Jr., *USA Today,* July 15, 1985.

[7] Unpublished draft report by the Policy Study Group on Terrorism of the New York State Division of Criminal Justice Services, 1984.

[8] *Wall Street Journal,* July 26, 1984.

[9] *CTT,* Vol. IX, Issue 2, p. 6.

pretensions" and lacking a true commitment to achieving political change by whatever methods were necessary (i.e., violence).[10]

The inability to conclude an agreement eventually spawned a series of internal disputes within the Weather Underground which finally compelled one faction to break away from the aboveground support apparatus (known as the Prairie Fire Organizing Committee) in 1975 and form a new group called the May 19th Coalition (sometimes referred to as the May 19th Communist Organization).[11]

The May 19th Coalition perceived its mission as the development and strengthening of links between politically militant blacks and Hispanics and their white ideological counterparts. It has largely become the aboveground political support apparatus of the RATF and the BLA. Its spokesperson, Judith A. Clark, participated in the Brinks robbery and is presently serving a sentence in New York's Woodbourne penitentiary. Clark is also alleged to have represented the May 19th Coalition at a conference sponsored by the PLO in Beirut, Lebanon, a month before the bank robbery.[12]

The May 19th Coalition actively pursues its aims in this country's prisons, recruiting black criminals in the prisons and recently paroled convicts. The Coalition typically establishes contact by offering black inmates free legal services and counseling. The prisoners who accept this offer are gradually drawn into "consciousness raising" meetings and seminars run by other inmates, where they receive rudimentary political indoctrination into the terrorists' ideological beliefs. Those who start to recognize their "victimization" by a "racist" and "unjust" society are earmarked for recruitment into the RATF or BLA upon their release.[13]

The May 19th cadres also function as couriers for the RATF and BLA, running a tight clandestine communications network among the imprisoned terrorists and their peers on the outside. This has enabled the BLA to maintain cohesion. The RATF network appears to extend across the United States. Ties have been established with the Mississippi chapter of the Republic of New Afrika group (a militant black organization that advocates armed struggle to establish an independent black country in the American south[14]) and with the Wells Spring

---

[10]Arostegui, "Terrorism in the United States," pp. 1–2.

[11]Ibid. The name was chosen to commemorate the birth dates of North Vietnam's Communist leader and founder, Ho Chi Minh, and American Black Muslim leader Malcolm X.

[12]*CTT*, Vol. IX, Issue 2, p. 4.

[13]Arostegui, "Terrorism in the United States," p. 2.

[14]The Republic of New Afrika has also been linked to a newer militant black terrorist organization called the New Afrikan Freedom Fighters (NAFF). Nine members of the

Commune (a group of paroled black prisoners operating in the San Francisco area). A van used in the Brinks robbery was traced to the Republic of New Afrika group, while an RATF member who participated in the robbery reportedly was a guest of the Wells Spring Commune during a visit to San Francisco. These groups share arms and intelligence, provide safehouses, and assist in the planning and execution of operations, including prison escapes. William Morales, an FALN bomber, was aided in his escape from prison in 1979 by the BLA and the May 19th Coalition. However, although the RATF supports the creation of an independent Puerto Rico, no formal cooperation between the Puerto Rican terrorists and the RATF has been detected beyond that of fraternal political support.[15]

There are believed to be fewer than 50 active RATF members.[16] However, the group is highly disciplined and its limited membership makes it virtually impossible to infiltrate. Evidence uncovered in connection with the Brinks holdup reflected the security precautions, organizational skill, logistical support, and operational expertise of the group.

In 1976, several small cells, each of which contained black, Hispanic, and white members, were established throughout the country. The New York cell, which participated in the Brinks' robbery, was headed largely by blacks but was supported by a "secondary team" of politically radical white women. The women transported messages among safehouses and cells, surreptitiously surveyed possible targets, and did odd jobs such as renting cars and apartments and buying weapons. The black members of the cell apparently carried out the terrorist operations themselves, although the women sometimes drove the getaway cars. Respectably dressed and looking like housewives or commuters, the women were an essential component of the cell's escape plans and operations.[17]

The New York cell is thought to have accumulated an estimated $1 million from bank robberies between 1976 and 1981. The money was used for safehouses, food, and other living expenses, as well as weapons and drugs; $100,000 was used to lease a three-story building in Harlem

---

NAFF were arrested by police in New York City on October 18, 1984, and were charged with planning the robbery of an armored car and the prison escape of a black radical. A 1984 Risks International report observed: "The breakup and identification of the NAFF confirms previous speculation about an expanding underground network concentrated in the New York area. The various cells appear to be self-sufficient but plugged into common above-ground support organizations." (*Quarterly Risk Assessment,* July-September 1984.)

[15] *Wall Street Journal,* July 26, 1984.

[16] *CTT,* Vol. IX, Issue 2, p.2.

[17] Ibid.; and *Wall Street Journal,* July 26, 1984.

that ostensibly served as an acupuncture clinic but in fact concealed an arms warehouse and a narcotics distribution center.[18]

When police raided a network of RATF safehouses in New York, they discovered detailed documents on several private companies and multinational corporations that were doing business with South Africa. It appears that the group was planning a terrorist campaign against American businesses involved in such dealings.

Apartheid in South Africa has long been an important issue among black militants and white activists in the United States, one that provides a common cause cementing the sometimes tenuous bonds that unite them. A few weeks before the Brinks robbery, a May 19th Coalition member had organized a violent demonstration at New York's JFK Airport to protest the arrival in the United States of the South African rugby team. During the team's nationwide tour, bombs exploded at the American Rugby Union offices in upstate New York and in Indiana. In December 1981, simultaneous bombings hit the South African Airways freight offices at JFK and an IBM office in Harrison, New York. The bombings were claimed by the hitherto unknown United Freedom Fighters, a name that may have been just a cover for RATF operations against or concerning South Africa. IBM was claimed to have been bombed because of its extensive investments in that country.[19]

Exactly a month later, a bomb exploded at the FBI office in New York. This attack was claimed by the Revolutionary Fighting Group, probably another pseudonym used by the RATF.[20] Terrorist groups often use different names in claiming responsibility for attacks, both to give an exaggerated impression of strength and numbers and to inhibit federal authorities from bringing charges against the terrorists. Under current laws, individuals linked to one group cannot be indicted for a bombing committed by another group unless they can be directly tied to that group—and this is not possible if the group exists only on paper.

On May 11, 1985, federal authorities arrested Marilyn Jean Buck, one of the Brinks robbery suspects who had been still at large. A search of Buck's apartment in Baltimore and other safehouses in Pennsylvania, New York, and Connecticut uncovered explosives, timers, weapons, cash, and false identity papers, and stolen vehicles were found in nearby garages. In addition, according to the FBI, "very

[18] *Wall Street Journal*, July 26, 1984; and Arostegui, "Terrorism in the United States," p. 4,

[19] Arostegui, "Terrorism in the United States," pp.2–3.

[20] Ibid., p. 3.

detailed" plans to bomb at least a dozen federal offices in Washington—including the Old Executive Office Building where Vice-President Bush's office is located and a building at the U.S. Naval Academy—were found in a file drawer marked "in progress." The evidence linked Buck and two accomplices to 16 bombings that had occurred in New York and Washington since 1982, including the bombing of the U.S. Capitol.[21]

UNITED FREEDOM FRONT

The United Freedom Front (UFF), a group apparently formed to protest U.S. involvement in Central America (not to be confused with the United Freedom Fighters), first came to public attention on May 12, 1983, when it claimed responsibility for bombing an Army Reserve Center in Uniondale, New York. The following night, the group struck again, bombing the Naval Reserve Center in Queens. Two more bombings of military facilities were carried out in 1983, one against a National Guard Armory in the Bronx and the other at a Navy Recruiting Center in East Meadow. In December 1983, the group claimed responsibility for two bombs that exploded in front of the Honeywell offices in Queens. A caller stated that the Honeywell bombs were set off to protest U.S. policy in Central America.

On January 12, 1984, a spokesperson for the UFF called UPI and directed federal agents to a mailbox in Brooklyn, which contained a communiqué that explained the group's mission to "protest U.S. imperialism around the world" and promised future bombings.[22] The Pentagon later warned 22 top defense contractors about the possibility of bombing attacks by terrorist groups.[23] On March 19, 1984, the UFF exploded its tenth bomb in the New York metropolitan area, this one at an IBM facility in the town of Purchase. This was the third attack by the group against IBM. Literature was left at the site of the blast, decrying the relationship between the United States and South Africa. The most recent operation claimed by the group took place on Septem-

[21]*The New York Times,* June 2, 1985.

[22]*CTT,* Vol. X, Issue 1, p. 3.

[23]The contractors were Boeing Military Airplane Co., Witchita, Kansas; Hughes Aircraft plants in Tucson and Los Angeles; the Rockwell Corp., Anaheim, California; Northrop, Los Angeles; Douglas and Rockwell Rocketdyne, Los Angeles; U.S. Air Force Plant #42, Palmdale, California; Lockheed, Sunnyvale, California; Martin-Marietta, Denver; Pratt and Whitney plants in East Hartford, Connecticut, and West Palm Beach, Florida; Lockheed-Georgia, Marietta, Georgia; Avco Systems Division, Wilmington, Massachusetts; and Westinghouse, Baltimore, Maryland. Plants operated by Fairchild Republic, General Electric, General Dynamics, and Thiokol were also warned.

ber 27, 1984, when the offices of Union Carbide in Tarrytown, New York, were bombed. An FBI agent explained that the group was attacking "what they perceived as the injustice of imperialism around the world and specifically Union Carbide for business dealings in South Africa."[24] Less than 24 hours earlier, a group calling itself Guerrilla Resistance took credit for a bomb that wrecked part of the South African consulate in New York City.

On March 13, 1985, seven members of the UFF were indicted on charges of bombing a total of ten businesses and military installations in the New York area. Five of the defendants had been arrested in Ohio in November 1984, including Raymond Luc Levasseur, a member of the Sam Melville-Jonathan Jackson Unit,[25] who was wanted for the 1981 murder of a New Jersey state trooper. The Sam Melville-Jonathan Jackson Unit and the United Freedom Front are now believed to have been the same group.

THE ARMED RESISTANCE UNIT

On November 6, 1983, the Armed Resistance Unit (ARU) claimed credit for the bomb that exploded in the Senate wing of the U.S. Capitol. An anonymous telephone caller explained that the blast was in retaliation for the U.S. invasion of Grenada a few days earlier and for the American presence in Lebanon. Investigations of the bombing and of two other Washington-area attacks against U.S. Navy targets, also claimed by the ARU, established a number of similarities between those operations and the bombings perpetrated by the UFF.[26] An FBI spokesman stated, "There is a possibility that the ARU and the UFF are one and the same. The communiqués issued by the ARU had language very similar to the communiqués issued by the UFF. Furthermore, there is consistency in their target selection and bomb construction."[27]

Speaking at a press conference shortly after the Capitol bombing, FBI Director William Webster said that the bomb detonated at the Capitol had a dual firing mechanism similar to the devices used in 10 or 11 other bombings during the previous two years. However, he

---

[24]*The New York Times,* September 28, 1984.

[25]This group was formed in the mid-1970s in the New England area. It took its name from Sam Melville, a white 1960s radical who was killed during the uprising in Attica Prison (New York) in 1971 and Jonathan Jackson, who was killed in a 1970 shootout with police after kidnapping a judge. The group is closely linked to the May 19th Coalition and the BLA.

[26]*Washington Times,* June 26, 1984.

[27]*CTT,* Vol. X, Issue 1, p. 3.

conceded that the FBI had no basis for concluding that all of the bombings came from "the same source." Webster noted that calls from people claiming responsibility for the bombing contained "the same type of words we experienced with the older dissident groups that have become dormant and which in a way have metamorphosed through other organizations we've been watching closely in light of the Nyack [Brinks] robbery." He emphasized that "the inferences we are drawing result not only from the rhetoric of claims but also the nature of the explosives which were used in all of these cases and the manner in which they were put together."[28]

In any event, the UFF bombings motivated many corporations to bolster security. The Grumman Corporation, for example, built a 3-ft-high concrete wall around its Bethpage, Long Island, plants to protect against truck or car bombs like those used in the Middle East. Moreover, federal bomb experts have been warning suburban police forces around the New York metropolitan area that they may be the next targets of terrorists. As Arostegui notes:

> At a time when security has been tightened around corporate and governmental facilities located in inner city areas, the suburbs have also come to offer softer targets for terrorists. During the past ten years many large corporations have moved their headquarters and main offices to suburban areas for the purpose of minimizing the risk of terrorist or criminal harassment. Corporate security directors admit that such facilities in the suburbs generally [have] less security than those in the inner city.[29]

THE LEFT-WING TERRORIST THREAT TO U.S. NUCLEAR WEAPONS PROGRAMS

Left-wing terrorist organizations have operated in the United States for nearly 20 years. The new leftist groups that have energed in recent years have embraced controversial, popular causes such as U.S. involvement in Central America and South Africa's apartheid policy in an attempt to broaden their base of support. Because nuclear disarmament is one of the most emotional and controversial domestic issues, it is not inconceivable that these groups might turn their attention to nuclear facilities and weapons.

However, no attempt has ever been mounted against any target as well-defended as a nuclear weapons facility. Left-wing terrorist operations have been confined to bombings of symbolic targets and armed

[28]Ibid., p. 5.

[29]Arostegui, "Terrorism in the United States," p. 4.

robbery. Bombings ("armed propaganda") are calculated to appeal to the terrorists' perceived constituency—leftward-leaning, politically concerned, nonviolent activists. Almost all the bombings occur at night and in public areas; they are planned to avoid inflicting casualties or widespread destruction; they are announced ahead of time, to allow the evacuation of the target site.

This does not imply that U.S. left-wing terrorists may not attack defended targets in the future. Clearly, few terrorist acts would be more symbolic than an attack on a nuclear weapons facility or the detonation of a small nuclear explosive. However, on the basis of past modus operandi, targeting patterns, and the general mindset of particular terrorist organizations, leftist terrorists seem unlikely to drastically alter their strategy or tactics.

Admittedly, as noted above, there is evidence that at least one left-wing group, the Sam Melville-Jonathan Jackson Unit, has entertained the idea of bombing a DOE weapons research laboratory. And the June 1983 Prairie Fire Organizing Committee book[30] of communiqués issued over the years by the Red Army Faction (RAF) in West Germany, *War on the Warmakers,* stated:

> We believe that the experience of the movement against imperialist war in West Germany is of great importance to all people who want to build an effective struggle against imperialism inside the U.S. As war grows, the limitations of the mainstream leadership of the anti-nuke, anti-war organizations are becoming obvious . . ..
>
> We support the people in the anti-nuclear movement who are taking up solidarity with El Salvador and are beginning to address the broader issue of U.S. Militarism. This process needs to be advanced and the movement must become more militant.[31]

However, the terrorists' attempt to link their goals and *raison d'être* to broader political issues is less a direct threat to the security of U.S. nuclear weapons facilities than a public relations gambit. Of course, terrorist operations could escalate to include nuclear targets, but this seems unlikely, since an assault on a nuclear site is inconsistent with the basic beliefs of most left-wing terrorists. One disquieting possibility, however, is that of some foreign power enlisting indigenous left-wing terrorists to carry out an operation against a nuclear weapons facility.

Because many members of militant black and radical white organizations have been arrested during the past two years, the level of

[30] *War on the Warmakers,* John Brown Book Club, 1983.

[31] IACP, *Terrorist Trends,* p. 11.

terrorism carried out by leftist groups may in fact have peaked. Only one recent act of violent terrorism has been linked to the left: the bombing of the Police Benevolent Association in New York City on February 23, 1985. The attack was claimed by the Red Guerrilla Defense, which had taken responsibility for the April 1984 bombing of the Israeli Aircraft Industries offices in New York and an explosion five months later at the South African consulate in New York.[32]

The RATF was effectively neutralized following the abortive holdup of an armored car in 1981 and has suffered further setbacks since then as the group's remaining members have been rounded up by police. Accordingly, it does not pose a credible threat to nuclear weapons facilities. Although the RATF is skilled in clandestine tradecraft, the use of disguises, and the use of firearms and explosives, its operational capacity is severely limited. The group has never attempted to attack a well-defended target, and although members have been involved in shootouts with police, they have generally tried to avoid armed confrontation (unlike right-wing terrorists).

Although on ideological grounds, the UFF would be among the groups most likely to attack a nuclear target, the modus operandi of the group suggests that such an operation would be symbolic rather than any sort of assault.

Left-wing terrorist groups in the United States have long found it difficult to recruit succeeding generations of terrorists. They have had to rely on the same "hard core" of individuals, many of whom went underground more than a decade ago. Accordingly, unless these groups are able to reverse their dependence on the old guard and recruit a new generation of members, it appears that they will continue to lose strength and momentum, as the few remaining terrorists on the outside become increasingly preoccupied with avoiding arrest rather than with staging new operations.

---

[32]Risks International, *Weekly Risk Assessment,* Vol. 2, No. 9, March 1, 1985.

V. RIGHT-WING TERRORIST GROUPS

> I personally regard them as more dangerous than the Klan groups from which they emanated.
>
> —William Webster, Director, FBI[1]

In recent years, evidence has repeatedly come to light of a well-organized network of extremist right-wing groups in the United States, all connected in some way to the so-called Christian Identity movement, which espouses a variety of anti-Semitic, racist, Christian-fundamentalist and anti-federalist beliefs. The basic tenets of the Identity movement, according to a report by the Anti-Defamation League of B'nai B'rith,

> include the doctrine of the inherent superiority of the white race, particularly the people of Northern European stock; the notion that those Nordic people are the descendents of the Biblical Israelites, and the necessary corollary that contemporary Jews are not. Vicious hostility toward non-white races and relentless vilification of Jews are further characteristics of this movement's theology of hate.[2]

Organized hate groups such as the Ku Klux Klan and various incarnations of Hitler's National Socialist (Nazi) party have existed in this country for decades. But the advent of extremist paramilitary groups oriented toward "survivalism," outdoors skills, guerrilla training, and outright sedition is a newer phenomenon. The geographical center of the movement is in the West, Midwest, and South, but the extremists' network extends as far east as North Carolina and Georgia, west to California and the Pacific northwest states, and on a north-south axis embracing Texas and Canada. The membership of these groups, according to Leonard Zeskind, research director of the Center for Democratic Renewal (a private research organization that monitors domestic right-wing activity), is estimated to be between 2,000 and 5,000 "hard-core activists" and between 14,000 and 50,000 sympathizers.[3] The Aryan Nations group, based in Hayden Lake, Idaho, serves

---

[1] *The New York Times,* April 4, 1985.

[2] Anti-Defamation League of B'nai B'rith, *Hate Groups in America: A Record of Bigotry and Violence,* 1982, p. 57.

[3] *The New York Times,* September 21, 1985. Risks International cites an estimated membership of approximately 4,000 persons. (See *Weekly Risk Assessment,* Vol. 2, No. 38, September 20, 1985.)

as the movement's "umbrella organization." The major right-wing terrorist groups are shown in Fig. 2.

Unlike most of their leftist counterparts, the members of these groups are, for the most part, not full-time terrorists.[4] Rather, they see themselves as "minutemen," the inheritors of the tradition of the American Revolution's Minutemen, who were available at a moment's notice to fight for their inalienable rights. They are bound together by their shared hostility to any form of government above the county level; their vilification of Jews and non-whites as children of Satan; obsession with achieving the religious and racial purification of the United States; belief in a conspiracy theory of powerful Jewish interests controlling the government, the banks, and the media; and advocacy of the overthrow of the U.S. government, or the "Zionist Occupation Government," as they call it. Some of the most fanatical members believe in the possibility of actually overthrowing the U.S. government and unleashing a nuclear attack on Israel. As far from reality as this may seem, the latter intention makes the rightists at

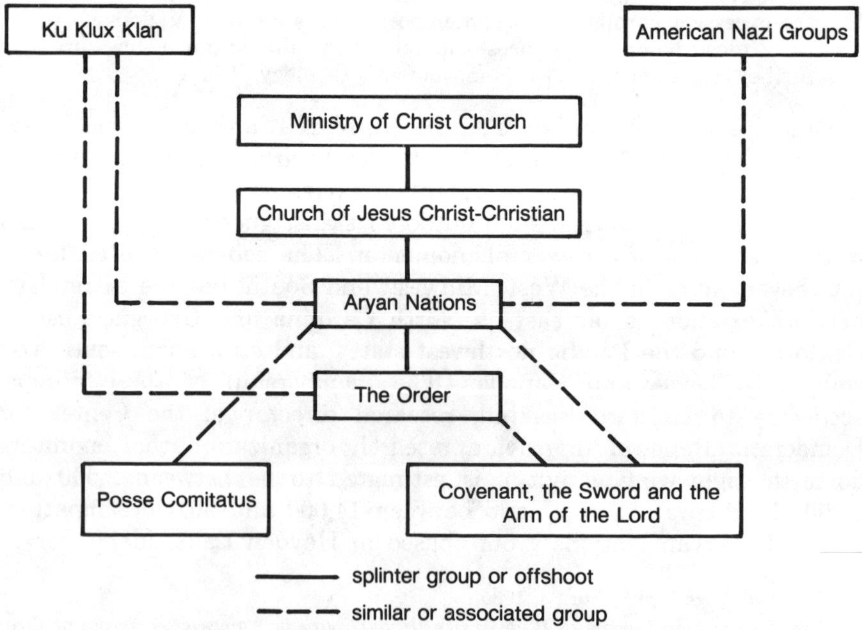

Fig. 2—Major right-wing terrorist groups

[4]The Order, which is discussed in detail below, appears to be an exception.

least potential adversaries of nuclear weapons sites from which they might attempt to steal material.

THE ARYAN NATIONS

The Aryan Nations has been described as an "extremist, anti-Semitic, Neo-Nazi group . . . [embracing] white supremacists, survivalists, militant tax resisters and Neo-Nazis."[5] Its headquarters is on a secluded and fenced 40-acre site at the edge of the Coeur d'Alene National Forest in Hayden Lake, Idaho. The group was founded in 1974 by Richard Butler, a 65-year-old former aeronautical engineer from California who moved to Idaho in 1973. Butler is also head of the Church of Jesus Christ-Christian, which also has its headquarters on the Aryan Nations compound. The Church is based on the white supremacist dogma espoused by a Depression-era anti-Semite, Dr. Wesley Swift. Butler became a member of Swift's Lancaster, California, congregation in 1963 and upon Swift's death in 1970 claimed leadership of the church for himself. The Church closely follows the aggressive anti-Semitic beliefs inherent in the Christian Identity Movement. The religious spearhead of the Identity Movement is the Ministry of Christ Church led by a retired U.S. Army colonel and expert in guerrilla warfare, the Reverend William Potter Gale of Mariposa, California, who, in fact, first introduced Butler to the Identity Movement. Butler named his ministry the Church of Jesus Christ-Christian in specific reference to the Identity Movement's contention that Jesus Christ was not a Jew, but an Aryan. This reflects their belief that the Lost Tribes of Israel are not composed of Jews but of "blue-eyed Aryans," that white Anglo-Saxons and not the Jews are the "Chosen People," and that the United States is the "Promised Land." In this view, Jews are viewed as impostors and children of Satan who therefore must be exterminated.[6]

The exact size of the Aryan Nations organization is not known, but Butler claims he has 6,000 members in the 50 states and Canada, as well as 400 regular subscribers to his mail-order sermons and 300 local congregationalists.[7] The Anti-Defamation League (ADL) argues that no more than 30 or 40 people regularly attend Butler's church and less

[5]IACP, *Terrorist Trends,* p. 13.

[6]Anti-Defamation League, *Hate Groups in America,* p. 54–55 and IACP, *Terrorist Trends,* p. 13.

[7]Ibid., p. 56.

than 200 persons come to Aryan Nations' rallies.[8] The International Association of Chiefs of Police (IACP) concedes that although there are Aryan Nations branches in 18 states, active membership is probably below 150.[9] Finally, observers of the group and unidentified informants from within it have estimated that the Aryan Nations has more than 500 active members.[10]

It is difficult to ascertain the Aryan Nations membership partly because it is an umbrella-type entity that serves as a centralized, coordinating body for a variety of different, but similarly oriented, organizations. Federal law enforcement officials believe that the Aryan Nations is "united more by ideology and theology than by organization and hierarchy."[11] One of the purposes of the group is to further the ties between members of various like-minded hate groups. Butler has described his efforts as the building of "an interrelationship of people with the same beliefs and ideas."[12] Ties have been forged with Canadian white supremacist groups and with West German neo-Nazi terrorists. Traudel Roeder, the wife of the leader of the West German neo-Nazi German Action Group, attended the annual Aryan Nations convention in 1982, "where she was honored with a special award."[13] The conference brought together some 300 representatives from 13 Ku Klux Klan, American Nazi, and other right-wing groups.[14]

Contact is maintained among these groups through a countrywide computer network ("bulletin board") based in Dallas that was created and is maintained by Louis Beam,[15] a former Grand Dragon (state leader) of the Texas chapter of the KKK who is described as the Aryan Nations' "ambassador-at-large."[16] Beam created the computer system after Canada refused to allow his racist literature into the country.[17] The "bulletin board," known as the Aryan Liberty Net, is self-described as the "pro-American, pro-White, anti-Communist network

[8]Ibid.

[9]IACP, *Trends In Terrorism,* p. 13.

[10]*The New York Times,* December 27, 1984.

[11]Ibid.

[12]Quoted in ADL, *Hate Groups in America,* p. 57.

[13]Ibid., p. 55. For details of European right-wing and neo-Nazi terrorist activities and the international links among them, see B. Hoffman, *Right-Wing Terrorism in Europe,* The Rand Corporation, N-1856-AF, March 1982; and B. Hoffman, "Right-Wing Terrorism in Europe Since 1980," *Orbis,* Spring 1984, pp. 16–27 (also issued as *Right-Wing Terrorism in Europe Since 1980,* The Rand Corporation, P-7029, October 1984).

[14]*The New York Times,* December 27, 1984.

[15]ADL, "Computerized Hate," Report of the Pacific Southwest Region, January-February-March 1985, p. 3.

[16]ADL, *Hate Groups in America,* p. 54.

[17]*Washington Post,* December 26, 1984.

of true believers who serve the one and only God—Jesus, the Christ . . . [and] is for Aryan patriots only." The network has one file entitled "Know Your Enemy," which contains the names, addresses, and phone numbers of members of the ADL and the U.S. Communist party, and so-called "race traitors" and "informers."[18] The "race traitors" category is broad enough to encompass federal judges, federal prosecutors, and Internal Revenue Service agents.[19] Access to the network is gained through passwords that interlock the various levels of secrecy of the system.

The ideology espoused by the the Aryan Nations is a mixture of racist and seditious dicta. In a 1982 newsletter, the group proclaimed its desire to "make it clear to ourselves and our enemies what we intend to do: We will have a national racial state at whatever price is necessary. Just as our forefathers purchased their freedom in blood, so must we. We have to kill the bastards."[20] The statement goes on to decry "the leadership of malicious, bastardizing politicians . . . [in] modern, decadent America [where] millions of whites watch in abject dismay and hopelessness as their great culture, heritage and civilization evaporates in the steaming, stinking, seething milieu of so many alien races, cultures and gods."[21] A more recent elucidation of the movement's goals, offered by the leader of the Midwest branch of the Aryan Nations, suggests future escalation of white supremacist terrorism:

> Soon, our own version of the "Troubles" will be widespread. The pattern of operations of the I.R.A. [Irish Republican Army] will be seen across this land.
>
> We, the older and less active spokesmen for the folks and faith, are being replaced by the young lions. These dragons of God have no time for pamphlets, for speeches, for gatherings; they know their role. They know their duty. They are the armed party which is being born out of the inability of white male youths to be heard. They are the products of the failure of this Satanic, anti-white federal monstrosity to listen to more peaceful voices, such as our own. . . . And now, as we had warned . . . once again the giants gather. Soon America becomes Ireland recreated.[22]

---

[18]*Washington Post,* December 15, 1984; and *The New York Times,* February 15, 1985.

[19]IACP, *Terrorist Trends,* p. 14.

[20]Ibid.

[21]Aryan Nations Newsletter No. 42, 1982, quoted in ADL, *Hate Groups in America,* p. 58.

[22]David Ausley, "Posse Comitatus: An Extremist Tax Protest Group," *TVI Journal,* Vol. 6, No. 1, Summer 1985, pp. 15–16.

The Aryan Nations exercises overall responsibility for the recruitment of new members to the white supremacist cause.[23] It is particularly active in U.S. prisons. Butler focuses his attention on penitentiaries in the West and South, where gangs known as the Aryan Brotherhood are common. Upon their release from prison, the ex-convicts are brought up to Hayden Lake to become members of the "Aryan Nations uniformed headquarters division." However, these new and younger members, according to Butler, created a schism within the Aryan Nations organization. "In their view," Butler explained, "the program . . . to reach the masses of our people through truth, logic, and reasonableness would never be allowed because the media are controlled by our enemies."[24] They therefore formed their own group within the Aryan Nations fold, which they called the *Bruder Schweigen* (Silent Brotherhood), or simply, The Order.

THE ORDER

The Order was founded in 1983 by 31-year-old Robert Matthews to "pursue a more violent approach towards making the U.S. a pure white Christian country."[25] This, Matthews and his followers reasoned, could only be achieved through the violent overthrow of the U.S. government. Racist tracts, written by Matthews, promulgate the view that the U.S. government has been taken over by Jews who are the sons of Satan and must be exterminated in what he describes as "a racial and religious Armageddon." The U.S. government, which Matthews disparagingly refers to as as Z.O.G.—the Zionist Occupation Government—must be completely destroyed as well.[26]

The Order's strategy is derived from a novel by William Pierce entitled, *The Turner Diaries,* which is considered to be "the Bible of the group."[27] A paramilitary group in the novel, also called the Order, acquires automatic weapons and plots and carries out terrorist attacks, assassinations of public officials, and bombings of public utilities. Several prominent Jews are murdered, the FBI headquarters in Washington, D.C., is bombed, and the U.S. nuclear arsenal is seized. A mass attack is then launched on Israel with nuclear weapons that obliterate

---

[23]*The New York Times,* October 20, 1985.

[24]*Washington Post,* December 26, 1984.

[25]Ibid.

[26]Ibid.

[27]*The New York Times,* December 27, 1984.

the Jewish state.[28] Because of The Order's strong identification with the fictional events in *The Turner Diaries,* this group would be the most likely of the right-wing extremists to attempt an assault on a nuclear weapons facility.

In June 1983, 17 members of The Order ambushed a Brinks armored car near Ukiah, California, making off with $3.6 million. Later that year, the group robbed a Continental armored car in Seattle of $500,000. Both robberies were executed in sophisticated "paramilitary" style: the robbers carried automatic weapons and displayed identical signs stating: "Get out or you die."[29] Matthews singlehandedly robbed a Seattle bank of $25,000. Through these holdups, the group amassed more than $4 million in $100 and $20 bills.[30]

One of the Order's declared aims is the destruction of utilities and communications networks. Although, to our knowledge, no operations of this kind have ever been attempted, during arrests of group members in 1984, the plans for a power dam on the Pend Oreille River in northeastern Washington were found. There was, however, no indication that the group had made serious plans to attack the dam.[31]

The Order regards all nonwhites as "mud people," and its members swear an oath to eliminate "their enemies," i.e., anyone "of Jewish race and white traitors who were cohorts [with] the Jews in trying to destroy our race."[32] Each member of the group can become an "Aryan Warrior" by accruing points based on the murder of Jews, blacks, federal judges, and FBI agents.[33] The Order has also "targeted for assassination federal, state and local authorities, members of Congress, judges, law enforcement officials, journalists, local politicians and the President of the United States."[34] In June 1984, members of the Order murdered Alan Berg, a Jewish radio talk-show host and an acerbic critic of the right wing. Berg was shot to death with a MAC-11 submachinegun (one of the rightists' favorite weapons) that had been altered to fire on full-automatic and was equipped with a silencer.

Matthews was killed in December 1984, after police and federal agents surrounded the cottage that he had barricaded himself in on

[28]*Washington Post,* December 26, 1984.

[29]*The New York Times,* December 27, 1984.

[30]*Washington Post,* December 26, 1984.

[31]*New York Times,* December 27, 1984.

[32]Court testimony of Order member Denver Daw Parmenter II, quoted in *The New York Times,* September 17, 1985.

[33]*The New York Times,* April 12, 1985.

[34]IACP, *Terrorist Trends,* p. 14. Senator Bob Packwood of Oregon has received threats against himself and his family from persons purporting to represent the group. (See *Executive Risk Assessment,* Vol. 7 No. 2, February 1985.)

Whidbey Island in Puget Sound, Washington, and four accomplices were arrested at the scene. When the cottage was searched, machine-guns, handguns, rifles, hand grenades, explosives, ammunition, false identification papers, and $64,000 in cash were discovered. In addition, police found a "declaration of war" against the U.S. government signed by group members, who referred to themselves as the "Aryan Resistance Movement."[35]

The Washington shootout led to a conference, held in Seattle in late December 1984, of U.S. federal attorneys from Alabama, California, Colorado, Idaho, and Oregon to work out a common strategy to coordinate their efforts against the rightists.[36] Within the next six months, 23 members of the Order were apprehended in 12 states. Information that has emerged from the trial of 11 of those arrested has thrown new light on the group's activities, goals, and ideology. All of the defendants have pleaded guilty to charges of participating in large-scale counterfeiting operations, armored car robberies, the firebombing of a cinema showing pornographic films in Seattle and a Jewish synagogue in Boise, Idaho, the murder of Alan Berg, the execution of Walter West (a suspected informant within the group),[37] and interstate transportation of stolen money.

Evidence presented at the trial also detailed the Order's plans to assassinate public officials, judges, and the heads of the three major American television networks, among others. The three network heads had been singled out, according to one of the defendants, because "the news media was responsible for indoctrinating our race [and] poisoning their minds."[39]

A plot to kill Baron Rothschild illustrates the lengths to which Order members are apparently willing to go to eliminate their enemies. It involved a planned suicide attack like those staged by Shi'a fanatics in the Middle East. Six members of the group planned to "blow . . . up the reception area" of the Seattle hotel where the Baron was staying. The attack, however, was never mounted because insufficient quanti-

---

[35] *The New York Times,* December 27, 1985.

[36] Ibid.

[37] West's elimination illustrates the almost paranoiac obsession with secrecy within the Order. West's "crime" was telling some members of the Aryan Nations Church about group plans. Another Order member was rebuked for cocaine use and for telling a friend about an armed robbery. The Order's security chief argued that he either be killed on the spot or sent on a suicide mission. In the end, however, he was fined and put on probation. (*The New York Times,* October 15, 1985.)

[38] *The New York Times,* September 13, 1985, and September 17, 1985.

[39] *The New York Times,* September 14, 1985.

ties of explosives were on hand at such short notice and there was not enough time for adequate planning.[40]

The Order's major goal was to create a "war chest" to support Identity movement activities throughout the country. Armed robbery and counterfeiting were to be the principal vehicles for raising the money. Federal authorities contend that more than $750,000 was distributed by the Order to other white supremacist groups. Recipients of this largess included Glenn Miller, a North Carolina KKK leader, who was given $300,000; Thomas Metzger, a former Klan leader and Congressional candidate who presently heads the White People's Political Association, who received $250,000; the National Alliance in Arlington, Virginia, led by William Pierce, which received $50,000; and Richard Girnt Butler, the leader of the Aryan Nations, who was given $40,000. Each member of the Order reportedly receives a salary of $400 per week; in addition, the seven members who participated in the armored-car robbery were allowed to divide $20,000 of the haul among themselves.[41]

The death of Matthews and the incarceration of the apparent nucleus of the group does not mean that the Order has been completely neutralized. Throughout 1984 and 1985, members of the Order have worked to establish a network of safehouses and support cells in the southwestern and southeastern United States.[42] Also, the identities of many members of the organization are not known. Members are all required to carry false identification (taken from headstones at local cemeteries) at all times and to use disguises and codenames to maintain secrecy,[43] and the group has regularly used voice-stress analyzers to detect informants and potential security risks.[44]

The damage to the Order caused by the death of Matthews and the mass arrests may not have been fatal, and a well-organized terrorist infrastructure and support network may still be in place.

THE COVENANT, THE SWORD AND THE ARM OF THE LORD

The Covenant, the Sword and the Arm of the Lord (CSA) was founded nearly a decade ago, but it only recently gained widespread public attention as a result of a police shootout involving two members of the Order, in which one trooper was killed and another was

[40]Ibid.

[41]*The New York Times,* September 13, 1985, and October 3, 1985.

[42]IACP, *Terrorist Trends,* p.14.

[43]*The New York Times,* October 15, 1985.

[44]*The New York Times,* October 15, 1985.

wounded. The gunman was found five days later near the CSA compound in Mountain Home, Arkansas. The authorities demanded to be allowed to enter the compound to search for weapons and other fugitives. When police raided the site, they uncovered an arsenal that included a modified MAC-10 submachinegun (similar to those used in the killings of the state trooper and Alan Berg), assorted pistols and rifles, 15,000 rounds of ammunition, materials used in the manufacture of hand grenades, an anti-tank rocket, plastic explosives, and a quantity of other explosives. They also found the "remnants of a minefield," a simulated paramilitary training area called "Silhouette City" (complete with cutout figures of policemen with Stars of David painted on their chests), an armored car under construction, and what one federal agent described as "a small but very efficient bomb factory," as well as computer and radio equipment and neo-Nazi hate literature. James Ellison and four members of the Order were arrested. This, according to authorities "proved a definite link between the Order and the Covenant." It was also determined that CSA members had undergone paramilitary training at the compound earlier.[45]

The search of the CSA compound provided further evidence of the geographical dimensions of the Identity movement. The CSA considers itself the paramilitary arm of the Church of Zarepath-Horeb (the name of a Biblical "purging place"), founded by Ellison, and its purpose is "to build an Ark for God's people during the coming tribulations on the earth." Accordingly, the 100 or so men, women, and children living in the compound prepare themselves for the coming Armageddon by stockpiling weapons, food, and valuables, and training in survivalist techniques and guerrilla warfare. The CSA journal describes the impending apocalypse as a limited nuclear war launched against the United States by "Russia and possibly China and Japan."[46]

POSSE COMITATUS

The group known variously as Posse by Law of Posse Comitatus, the Sheriff's Posse Comitatus, or simply as Posse Comitatus, takes its name from the Latin phrase meaning "Power of the County."[47] It was founded in 1969 by Henry Lamont (Mike) Beach,[48] one of this country's older Nazis, who allegedly belonged to the U.S.-based pro-

[45] *The New York Times,* April 23 and April 28, 1985.

[46] ADL, *Hate Groups in America,* p. 51.

[47] Ibid., 4 June 1983.

[48] See Phillip Finch, "Renegade Justice," *The New Republic,* April 25, 1983, p. 10.

Nazi Silver Shirts during the 1930s.[49] Throughout the 1970s, local chapters of the organization were founded in almost every state in the country. According to one observer, "The group maintained a low profile, concentrating on passive tax protest activities, such as providing legal counsel for tax evaders and prompting constitutional justification for local county rules and powers." However, during "the last two years, Posse Comitatus has become more violent, particularly in the midwest and far northwest."[50] To some extent, this trend has been fueled by the plight of financially depressed farmers in the West and Midwest. Searching for a reason for their economic predicament, some farmers have accepted facile political and religious interpretations that blame their situation on grandiose conspiracy theories of secret cabals composed of Jews, bankers, and the federal government.[51]

The basic *raison d'être* of the Posse Comitatus is anti-federalism. Its members are opposed to federal and state income tax, the existence of the Federal Reserve System, and the supremacy of the federal judiciary over local courts. They decry federal and state taxes in particular as "Communist and unconstitutional," and they advocate a return to the gold standard and abolition of the Federal Reserve Bank.

The group holds that county law is the only legitimate government authority and calls for the formation of "a posse in each county to assist the only legitimate law enforcement authority, the county sheriff, in combating the unlawful acts of others, particularly those of Federal and state officials."[52] But according to one observer,

> the Posse is not strictly a tax protest group. What passes for a world view among many of its members is the wilted right-wing conspiracy theory that a cabal of bankers and anonymous insiders is manipulating global events. The group's sole official policy is simply resistance to all state and national government. . . . They are hyper-libertarians, angry and antisocial.[53]

During the past two years, Posse members have attacked local, state, and federal law enforcement officers attempting to serve subpoenas. In addition, Posse adherents stopped for traffic violations refuse to present their driver's licenses or sign citations and will not register their vehicles with the state or purchase license plates. Members also

---

[49]*The New York Times,* June 11, 1983.

[50]Audsley, "Posse Comitatus," p. 13.

[51]*The New York Times,* August 17, 1985, September 20, 1985, and September 21, 1985.

[52]FBI, *FBI Analysis of Terrorist Incidents,* 1983, p. 27.

[53]Finch, "Renegade Justice," p. 10.

file "frivolous" and harassing lawsuits against arresting officers, IRS and ATF agents, and other law enforcement personnel.[54]

Although the exact size of the Posse Comitatus movement is not known, estimates of its membership range between 1,000 and 3,000 persons.[55] Leaders of the group, however, maintain that Posse Comitatus has chapters in every state but Hawaii and has 2,000 members in Wisconsin alone.[56] What is known is that members of the Posse are scattered around the West and Midwest and that "loosely organized state branches" exist in Oregon, California, Wisconsin, and Missouri.[57] Local chapters are controlled by independent leaders who maintain a strong if somewhat isolated profile in the local county or parish. The typical Posse cell consists of at least seven persons (between the ages of 18 and 45) who will, when ordered by the sheriff (often elected by Posse members), "uphold the laws and preserve the peace of the county's people."[58]

The organization holds "counterinsurgency seminars" throughout the United States, particularly in the West and Midwest. Kansas Attorney-General Robert Stephan described a typical seminar as a three-day event where approximately 50 participants are trained as "killer teams in hand-to-hand combat techniques, the administration of poisons, night combat patrol and murder by ambush." Other seminars allegedly have included instruction in bomb making. The Posse is reported to have also engaged "in joint military maneuvers" with other rightist groups. According to one federal agent, "There has been some information that they are into RPGs (rocket-propelled grenades), mortars, explosive and protective equipment, heavy-duty armor."[59]

The typical member of the Posse Comitatus is white, but perhaps not reflexively racist. He has deep, fundamentalist Christian views and is not likely to have finished high school. He is from a rural area where he has lived all his life, and he is probably self-employed in some manual labor.[60]

There have been recent reports of increased recruitment activity and propaganda efforts aimed at Midwestern farmers. KKK publications such as *The White Patriot,* the fledgling Populist Party's organ, *The Spotlight,* and the extremist National Agricultural Press Association's

[54]Audsley, "Posse Comitatus," pp. 13–14.

[55]Ibid., p. 9; and *The New York Times,* June 11, 1983.

[56]*The New York Times,* June 4, 1983.

[57]Finch, "Renegade Justice," p. 10.

[58]Audsley, "Posse Comitatus," p. 13.

[59]*The New York Times,* June 11, 1983.

[60]Finch, "Renegade Justice," p. 10.

(now discontinued) *Primrose and Cattleman's Gazette* have appealed to farmers to stop paying taxes, ignore state and federal judicial authority, abjure from using U.S. currency, and arm themselves against any attempt by government authorities to enforce these laws.

EMERGING LINKS BETWEEN WHITE SUPREMACISTS AND BLACK MUSLIMS

An ironic recent trend has been the forging of bonds between the white supremacists and Louis Farrakhan and the Nation of Islam. In October 1985, Tom Metzger, the leader of the California-based White People's Political Association, announced that he had led a "white nationalist" delegation (comprising nine members of his organization) at a Nation of Islam rally as guests of the Black Muslims. In addition, he had donated $100 on behalf of his group to the Nation of Islam. Contact had been established between the whites and blacks during Jesse Jackson's bid for the Democratic Party nomination for President in 1984, when the White People's Political Association, according to Metzger, furnished the Nation of Islam with information pertaining to "the movements of Jewish terrorist groups that could have caused damage to Jesse Jackson's campaign." The terrorist groups referred to were the JDL and the JDO. "We wanted to assure the Nation of Islam," he added, "that white nationalists generally would not be part of any attacks on the Jackson campaign, and that if it happened it would be a result of somebody else trying to make it look like us."[61]

The confluence of interests of right-wing white extremists and black radicals was evident at a white supremacist convention in the same month. In a speech delivered at the the convention, Metzger declared, "America is like a rotting carcass. The Jews are living off the carcass like the parasites they are. Farrakhan understands this."[62] Both movements agree that whites and blacks are separate nations and therefore should live apart in their own countries, under their own governments. Both ascribe all sorts of evil intentions and actions to Jews and argue that powerful Jewish interests have taken control of the country and are at the root of America's problems.

---

[61]*The New York Times,* October 3, 1985.

[62]*Los Angeles Times,* October 11, 1985.

ANTI-ABORTION TERRORISM

In recent years, some elements within the anti-abortion movement have turned to violence to end legalized abortion in the United States. Since March 1982, there have been 42 bombings of abortion clinics and more than 150 acts of related vandalism in Florida, California, Washington, Delaware, Maryland, Texas, Georgia, and the District of Columbia. More than 75 percent of the bombings have occurred during the past year, and 12 persons have been arrested in connection with these attacks. Police and federal law enforcement officials have stressed that there is no evidence of an organized network of local anti-abortion terrorist cells.[63]

Although the anti-abortion terrorists are disparate, they use similar names in claiming credit for their attacks, calling themselves the Army of God, or God's Army, or the Armies of the Living God. They are angered by the the failure of organized fundamentalist Christian groups, such as the Moral Majority, their representatives in Congress, and their religious and political leaders to enact legislation or a constitutional amendment banning abortions. They are primarily fundamentalist Christians and are sincerely disturbed by abortion.

Although organized anti-abortion political action groups and lobbies condemn the violence, they defend the motives of the bombers. Cal Thomas of the Moral Majority stated while he deplored the bombings, he recognized that throughout history "certain groups or individuals thought something dramatic was necessary to force lethargic political leaders off dead center so that they might focus on higher moral issues. . . . The violence is to be deplored . . . but if it can be used to translate the abortion issue into a national debate, then it's served its end."[64]

In addition to bombing and vandalism of abortion clinics, anti-abortionists have perpetrated one kidnapping and have fired a shot through the window of Supreme Court Justice Harry Blackmun's home. (Blackmun authored the Court's majority opinion on *Roe v. Wade* in favor of abortion in 1973.)

The anti-abortion movement is increasingly attracting persons from the fringes of the Identity movement. The Posse Comitatus, in fact, has stated that abortion is part of the global conspiracy "master-

[63]*The New York Times,* November 22, 1982. The FBI has, in fact, declared that since there is no evidence of a national conspiracy behind the attacks and no evidence that the bombings have violated civil rights or are the work of "terrorists," violence against abortion clinics is outside of FBI jurisdiction.

[64]James Ridgeway, "Unholy Terrorists", *Village Voice,* January 25, 1985.

minded" by the Jews.[65] The growing involvement of violent, well-armed right-wing groups could have a profound impact on the anti-abortionists' terrorist capabilities.

THE RIGHT-WING TERRORIST THREAT TO U.S. NUCLEAR WEAPONS PROGRAMS

Of the terrorist organizations active in this country, the right-wing extremists appear to pose the most serious threat to U.S. nuclear weapons facilities. They are serious in their beliefs and dedicated to their causes, and they are willing to use violence in pursuit of their goals. The neo-Nazis, anti-Semites, racists, and survivalists who comprise this movement are well-trained in the use of arms and explosives, skilled armorers and bombmakers, and adept at guerrilla warfare techniques. They are more violent than their leftist counterparts, but they are often similarly motivated by controversial, political issues (e.g., the abortion issue).

In terms of their potential threat to nuclear weapons facilities, two characteristics of right-wing terrorist groups are particularly important:

1. They have an apocalyptic vision of an impending Armageddon.
2. They have a visceral hatred of Jews, and the announced goal of some elements is to destroy Israel with nuclear weapons.

Because none of the extremist groups possess nuclear weapons, they could only accomplish the latter goal by stealing weapons or materials from U.S. nuclear arsenals. Thus, they represent at least an indirect threat to nuclear weapons sites.

The Order, because of its close identification with its fictitious namesake in *The Turner Diaries,* is probably the group most likely to attack a nuclear facility. Its has demonstrated its organizational and logistical capacity, its planning skills, and the firepower at its disposal. It allegedly planned one suicide attack that was to occur in a crowded hotel reception area. This operation, although never carried out, demonstrates the fanaticism of group members and the lengths to which they are prepared to go to achieve their aims and eliminate their "enemies."

The CSA, with its obsession with commando warfare and survivalist training and its avowed belief in the coming Armageddon, has a motive, and perhaps the means, to attack a nuclear weapons facility.

[65] Ridgeway, "Unholy Terrorists."

The Posse Comitatus poses a significantly smaller threat. The threat posed by anti-abortion terrorists to U.S. nuclear weapons programs is insignificant at best. Their narrow, specific cause, their lack of sophistication, and their lack of weapons capability make it highly unlikely that they would attack a nuclear weapons facility or even pose any serious threat to one. But some specific, contentious issue could arise to provoke any of the extremist groups to attempt some desperate operation. Further, we cannot exclude the possibility of some charismatic figure emerging who is able to unite the diffuse extremist movement and organize more ambitious operations.

Thus far, however, dangerous and tactically proficient as the right-wing extremists are, they have not shown themselves to be particularly innovative. For the time being at least, there seems to be little chance that they will attempt to acquire a nuclear weapon.

VI. CONCLUSION

The threat to U.S. nuclear weapons facilities posed by terrorist groups in the United States is not high at this time. Terrorist activity in this country has primarily consisted of symbolic bombings to call attention to political causes. Terrorists have avoided attacking *defended* sites, such as nuclear weapons facilities. However, we cannot reach the same conclusion with regard to possible state-sponsored terrorism in the United States. With their greater resources and lesser political concerns, state-sponsored terrorist groups could constitute a significant danger to nuclear weapons sites.

This is not to say that the threat from domestic terrorist groups is negligible. On the basis of past modus operandi, targeting, motivation, and mindset, Islamic fanatics, right-wing terrorists, left-wing terrorists, and Puerto Rican separatists could conceivably attack a nuclear installation. Less threat is posed by Jewish extremists, anti-Castro Cuban militants, Armenian radicals, and anti-abortionists.

The Islamic fanatics seem most likely to become state-sponsored. They have close ties with Libya and Iran, and Qadaffi has openly stated his desire to obtain a nuclear weapons capability. Right-wing terrorists might assault a weapons facility to obtain a nuclear device with which to hasten the Armageddon they see as inevitable, to use against Israel, or to destroy the U.S. government. A left-wing terrorist attack on a nuclear facility seems unlikely, since the leftists have consistently avoided operations that might cause widespread bloodshed and destruction and thereby alienate their constituency, both real and imagined. But it would not be out of character for them to undertake symbolic bombings of nuclear sites. Finally, the threat from the Puerto Rican separatists, other ethnic/émigré organizations, and anti-abortionists is also remote.

There are important distinctions among the goals and motivations of the three types of terrorist organizations that could affect their potential targeting of nuclear weapons facilities. Leftist groups might be expected to attack facilities to destroy weapons or to attract attention to the nuclear arms race and the need to rid the world of tools of mass destruction. Right-wing groups, on the other hand, might assault a facility to acquire nuclear weapons so that they could precipitate the global conflagration that they believe will result in the destruction of their enemies. Ethnic/émigré groups would probably have neither the motivation nor the inclination to attack a nuclear weapons facility unless they were hired by an outside patron-state or intent on blackmail.

Because of the extreme consequences that would result from *any* nuclear detonation, even the remotest likelihood of one cannot be dismissed as insignificant. In that respect, risk analysis of threats to nuclear weapons sites differs from other risk analyses: One is one too many. But a defense that would preclude every possible attack by every possible group for any possible motive is not even theoretically conceivable. Those charged with the security of nuclear weapons sites must be satisfied with doing the best they can, on the basis of the best and most complete available knowledge of all potential adversaries. The situation confronting the defenders is one of constant flux: Technology continues to improve, motivations change, new groups arise, old groups vanish, new modes of terrorism produce new risks, and the sensitivities of public opinion change in unpredictable ways. The defense must therefore be dynamic, to respond as effectively as possible under the most difficult circumstances.

BIBLIOGRAPHY

Books and Articles

Anti-Defamation League of B'nai B'rith. "Computerized Hate," Report of the Pacific Southwest Region, January-February-March 1985.

——, *Hate Groups in America: A Record of Bigotry and Violence,* 1982.

Arostegui, Martin C., "Terrorism in the United States," *Clandestine Tactics and Technology: Group and Area Studies,* Vol. IX (n.d.).

Audsley, David, "Posse Comitatus: An Extremist Tax Protest Group," *TVI Journal,* Vol. 6, No. 1, Summer 1985.

Cordes, Bonnie, "Armenian Terrorism in America," *TVI Journal,* Spring 1984.

——, et al., *Trends in International Terrorism, 1982 and 1983,* The Rand Corporation, R-3183-SL, August 1984.

Dolgin, Janet L., *Jewish Identity and the JDL,* Princeton: Princeton University Press, 1977.

Federal Bureau of Investigation, Terrorist Research and Analytical Center, *FBI Analysis of Terrorist Incidents in the United States,* Reports for 1981, 1982, 1983, and 1984 published annually in March of the succeeding year.

Finch, Phillip, "Renegade Justice," *The New Republic,* April 25, 1983.

Hoffman, Bruce, "The Cuban Anti-Castro Terrorist Movement," *TVI Journal,* Spring 1984.

——, "The Jewish Defense League," *TVI Journal,* Spring 1984.

——, *Right-Wing Terrorism in Europe,* The Rand Corporation, N-1856-AF, March 1982.

——, "Right-Wing Terrorism in Europe Since 1980," *Orbis,* Spring 1984 (also issued as *Right-Wing Terrorism in Europe Since 1980,* The Rand Corporation, P-7029, October 1984).

International Association of Chiefs of Police, Center for the Study of International Criminality, *Clandestine Tactics and Technology,* Vol. IX, Issue 2 (n.d.).

——, *Terrorist Trends: The Quarterly Intelligence Reporter,* 1985.

Porter, Jack Nusan, and Roy Delbyck, "The Jewish Defense League: New Tensions in the Old Melting Pot," *Christianity and Crisis,* April 15, 1972.

Risks International, Inc., *Executive Risk Assessment,* various issues.

——, *Weekly Risk Assessment,* various issues.

Sater, William, *Puerto Rican Terrorists: A Possible Threat to U.S. Energy Installations?* The Rand Corporation, N-1764-SL, October 1981.

——, "Violence and the Puerto Rican Separatist Movement," *TVI Journal,* Vol. 5, No. 1, Spring 1984.

Weiss, Ellen, "Islamic Terrorist Threat in U.S.," *TVI Journal,* Summer 1984.

Newspapers

Latin (Buenos Aires).
Los Angeles Herald Tribune.
Los Angeles Times.
Mamara (Istanbul).
Miami Herald.
New York Times.
United Press International Wire Service.
USA Today (Washington, D.C.).
Wall Street Journal (New York).
Washington Post.
Washington Times.